Nutritional Neurosciences

Series Editor
Mohamed Essa, Sultan Qaboos University, Qaboos, Oman

This book series aims to publish volumes focusing on both basic and clinical research in the field of nutritional neuroscience with a focus on delineating the effect of nutrition on brain function and behavior. The books will examine the role of different nutrients, food agents and supplements (both macro and micro) on brain health, neurodevelopment, neurochemistry, and behaviour. The books will examine the influence of diet, including phytochemicals, antioxidants, dietary supplements, food additives, and other nutrients on the physiology and metabolism of neurons, neurotransmitters and their receptors, cognition, behavior, and hormonal regulations.

The books will also cover the influence of nutrients and dietary supplements on the management of neurological disorders. It details the mechanism of action of phytonutrients on signaling pathways linked with protein folding, aggregation, and neuroinflammation. The books published in the series will be useful for neuroscientists, nutritionists, neurologists, psychiatrists, and those interested in preventive medicine.

Wael Mohamed • Rajat Sandhir
Editors

Trace Elements in Brain Health and Diseases

Springer

Editors
Wael Mohamed
Department of BMS
International Islamic University Malaysia
Kuantan, Malaysia

Rajat Sandhir
Department of Biochemistry
Punjab University
Chandigarh, India

ISSN 2730-6712 ISSN 2730-6720 (electronic)
Nutritional Neurosciences
ISBN 978-981-99-1512-5 ISBN 978-981-99-1513-2 (eBook)
https://doi.org/10.1007/978-981-99-1513-2

This Springer imprint is published by the registered company Springer Nature Singapore Pte Ltd.
The registered company address is: 152 Beach Road, #21-01/04 Gateway East, Singapore 189721, Singapore

It is with genuine gratitude and warm regards that I dedicate this work to Menoufia Medical School: Thank you so much for the unending financial and logistical support during my Ph.D. journey and afterward during my sabbatical leave. I could not do any of this without your love and glory.

Wael Mohamed

Preface

"To study the phenomenon of disease without books is to sail an uncharted sea, while to study books without patients is not to go to sea at all."
William Osler (1849-1919)

Recently, the impact of nutrition and food intake has been highly investigated to study its impact on our brain function and its development as it was shown that the diet we take will determine the outcome of certain brain disorders such as brain injury and stroke. Along with its effects on cardiovascular diseases and cancer development, nutrition and diet have been shown to be involved in preserving our mental cognitive function and behavior. Recent studies have implicated the development or exacerbation of certain neurological disorders to an imbalance in our nutritional intake and our diet, especially iron. The book *Trace Elements in Brain Health and Diseases* will be published by Springer Nature under the Nutritional Neurosciences series (https://www.springer.com/series/16639). The aim of this project is to assemble global perspectives concerning the relationship between trace elements and the brain in health and diseases. This book will be part of the Nutritional Neurosciences series that covers multiple domains within nutritional neuroscience. The inclusion of trace elements in this prestigious series will help introduce new readers to the subdiscipline and increase the number of global conversation partners.

In the human body, trace elements are most plentiful. It is well known that trace elements are an essential component of the body's metabolism. Previous studies have shown that deficiency of various trace elements in the brain triggers mental impairment in infants and young children, for example, verbal and body coordination delays and psychomotor disorders. Therefore, the mechanism and control of brain trace elements metabolism should be researched and understood thoroughly. On this basis, it is important to explore the relationship between brain trace elements control and the incidence of nervous system diseases and discover new metabolism-related therapeutic targets in order to break down the limitation of nervous system disease prevention and treatment.

Trace Elements in Brain Health and Diseases addresses cutting-edge areas of research of high significance for public health and translational medicine. The book discusses the comprehensive research history of trace elements and their significant role in the pathogenesis of central nervous system (CNS) diseases. The book also identifies how trace elements support function as well as the molecular mechanisms underlying their neuroprotectant activity. This topic is among the most interesting and challenging areas of contemporary translational biological and medical research, with implications for preventive and therapeutic approaches in age-related neurodegenerative disorders. This book explores the molecular mechanisms of brain trace elements including age-related metabolic pathways, mitochondrial nutrients, neurodegeneration and CNS disorders, cell signaling, and neuronal functions.

Coming from a background in the area of neuro-psychiatric health research, the editors (Drs. Mohammed and Sandhir) have decided to collaborate with other colleagues with expertise in areas of nutritional neuroscience to have a comprehensive book entitled *Trace Elements in Brain Health and Diseases,* which included eight chapters.

Overall, this new book provides updated and novel concepts in the field of neurological disorders and their relation to trace elements. The new compilation will be of high interest among researchers and clinical scientists involved in neuropsychiatry, nutrition, and biochemistry.

Finally, we thank all the authors for their significant effort in writing such excellent chapters for this new edition. We are also sincerely grateful to each author for their patience during the compilation and final editing of this book.

Kuantan, Malaysia Wael Mohamed
Chandigarh, India Rajat Sandhir

Acknowledgments

First, we would like to express our sincere appreciation to all the authors who contributed to this timely project. The high level of devotion and dedication between the authors and editors made writing this book an enjoyable journey. In addition, we also extend our gratitude to the authors who are in the fields of neurology and neuropsychiatric research for delivering years of their experience and work in different areas of psychiatric disorders to deliver such an elegant piece of work. The topics and applications discussed are of great value in the areas of nutrition, neurological disorders, and neurodegeneration. Finally, we would like to thank many of our friends and colleagues for their unconditional love, encouragement, and inspiration throughout the project. Thank you.

Contents

Editors and Contributors

About the Editors

Wael Mohamed, BBCH, MMSc, MD/PhD, PsyD is a Physician Neuroscientist. Dr. Mohamed received his PhD from PSU, USA, and is currently working as Professor Madya Dr. in IIUM Medical School, Malaysia. Dr. Mohamed has been invited to deliver more than 120 lectures locally and abroad. He published over 70 peer-reviewed papers related to neuroscience/psychiatry with an *h*-index of 18. Moreover, he is an editor of *Frontiers in Neurology* and *PLOS-ONE* journals and has edited many journal special issues on brain disorders. Additionally, he is editing a few books in the field of neuroscience with Springer, CRC, and Cambridge. He received many research grants from national and international organizations namely IBRO, ISN, MJF, STDF, FRGS, and INDO-ASEAN with a total research funding of half a million US$. He is now an active partner in GP2 consortium (IPDG-Asia).

Rajat Sandhir, MSc, PhD, FIAN, FABMS has received his MSc and PhD degrees in Biochemistry from the Postgraduate Institute of Medical Education and Research, Chandigarh. He has been at the Department of Biochemistry, Panjab University, for more than 20 years. He has 28 years of teaching and research experience. His research interests are to understand the biochemical and molecular mechanisms involved in the development of neurodegenerative conditions like metabolic encephalopathies, dementias and brain injury with a particular interest to investigate the role of oxidative stress, mitochondrial dysfunctions, and alterations in permeability of blood-brain barrier. He has an interest in nutritional neuroscience, particularly about the role of trace elements especially zinc and selenium in the brain. In addition, his interest is also to identify neuroprotective strategies that could ameliorate neurodegenerative conditions. He has over 200 papers to his credit and has mentored over three dozen students for PhD. He has an h-index of 49 and i10 index of 140. He has been on the editorial board of many international journals. He was awarded with "KT Shetty Memorial Oration (2021)" by the Indian Academy of Neurosciences. He has recently been conferred with "Mrs. Abida Mahdi

Award–2022" by the Indian Academy of Biomedical Sciences for outstanding contributions to the field of neurosciences. He is also a Fellow of Indian Academy of Neurosciences and Indian Association of Biomedical Scientists. He has been listed among the World 2% Scientists for the year 2020 released now by the Stanford researchers from Panjab University.

Contributors

Driss Ait Ali Biological and health sciences team, Higher Institute of Nursing Professions and Health Techniques, Ministry of Health, Ouarzazate, Morocco

Shi-Hui Cheng School of Biosciences, Faculty of Science and Engineering, University of Nottingham Malaysia, Semenyih, Selangor, Malaysia

Mario Eid Faculty of Medicine, Lebanese University, Hadath, Lebanon

Hafida El Ghachi Neurosciences, Pharmacology and Environment Team, Laboratory of Clinical, Experimental and Environmental Neurosciences, Faculty of Medicine and Pharmacy, Cadi Ayyad University, Marrakech, Morocco

Samer El Hayek Medical Department, Erada Center for Treatment and Rehab, Dubai, United Arab Emirates

Brahim El Houate Biological and health sciences team, Higher Institute of Nursing Professions and Health Techniques, Ministry of Health, Ouarzazate, Morocco

Abdelaati El Khiat Laboratory of Clinical and Experimental Neurosciences and Environment, Faculty of Medicine and Pharmacy, Cadi Ayyad University, Marrakech, Morocco

Biological and health sciences team, Higher Institute of Nursing Professions and Health Techniques, Ministry of Health, Ouarzazate, Morocco

Interdisciplinary Laboratory in Bio-Resources, Environment and Materials, Higher Normal School, Marrakech, Morocco

Mohamed El Koutbi Biological and health sciences team, Higher Institute of Nursing Professions and Health Techniques, Ministry of Health, Ouarzazate, Morocco

Bilal El-Mansoury Faculty of Sciences, Department of Biology, Chouaib Doukkali University, El Jadida, Morocco

Halima Gamrani Neurosciences, Pharmacology and Environment Team, Laboratory of Clinical, Experimental and Environmental Neurosciences, Faculty of Medicine and Pharmacy, Cadi Ayyad University, Marrakech, Morocco

Youssef Ait Hamdan Interdisciplinary Laboratory in Bio-Resources, Environment and Materials, Higher Normal School, Marrakech, Morocco

Ahmad Hassan Department of Psychiatry, Washington University School of Medicine, St. Louis, MO, USA

Sarah Hassan The Chicago School of Professional Psychology, Chicago, IL, USA

Nazeha A. Khalil Nutrition and Food Sciences Department, Faculty of Home Economics, Menoufia University, Shibin El Kom, Egypt

Najib Kissani Laboratory of Clinical and Experimental Neurosciences and Environment, Faculty of Medicine and Pharmacy, Cadi Ayyad University, Marrakech, Morocco

Abdelali Ben Maloui Neurosciences, Pharmacology and Environment Team, Laboratory of Clinical, Experimental and Environmental Neurosciences, Faculty of Medicine and Pharmacy, Cadi Ayyad University, Marrakech, Morocco

Wael Mohamed Department of Basic Medical Sciences, Kulliyyah of Medicine, International Islamic University Malaysia (IIUM), Kuantan, Malaysia

Clinical Pharmacology Department, Menoufia Medical School, Menoufia University, Menoufia, Egypt

Mjid Oukhrib Biology, Ecology and Health Unit, Department of Biology, Faculty of Science, Abdelmalek Essaadi University, Tetouan, Morocco

Neurosciences, Pharmacology and Environment Team, Laboratory of Clinical, Experimental and Environmental Neurosciences, Faculty of Medicine and Pharmacy, Cadi Ayyad University, Marrakech, Morocco

Al-Hassan Soliman Oral Biology Department, Faculty of Dentistry, Sinai University (SU), Arish, North Sinai Peninsula, Egypt

Lahcen Tamegart Biology, Ecology and Health Unit, Department of Biology, Faculty of Science, Abdelmalek Essaadi University, Tetouan, Morocco

Neurosciences, Pharmacology and Environment Team, Laboratory of Clinical, Experimental and Environmental Neurosciences, Faculty of Medicine and Pharmacy, Cadi Ayyad University, Marrakech, Morocco

Neda Valian Neuroscience Research Center, Shahid Beheshti University of Medical Sciences, Tehran, Iran

Ke Tian Yong School of Biosciences, Faculty of Science and Engineering, University of Nottingham Malaysia, Semenyih, Selangor, Malaysia

Chapter 1
Dietary Iron and Brain Development

Nazeha A. Khalil

Abstract Trace elements are very important micronutrients needed for human health. Iron either in heme (Fe+2) or nonheme (Fe+3) sources is one of the most effective elements required for different physiological activities within the neurodegenerative process/developments. Iron availability is depending on different factors such as the human health state (infection and inflammatory stress), nutrition, and dietary consumptions in addition to human ages and genders. Iron is vital for brain development with different critical functions either in low or high levels; iron homeostasis. The current chapter aimed to highlight the iron dysregulations with different relationship between iron status and neurological/brain developments or sensitive to iron-related neural changes; neurophysiological function with cognitive and motor or socio-emotional development. It has also been discussed the iron deficiency (ID) effects with different prevention strategies for protecting maternal health and brain developments especially at Alzheimer's disease (AD) and Parkinson disease (PD) conditions. Also, dietary antioxidant sources as polyphenols in tea could help with AD improvement because of their chelating iron, scavenging oxygen-free radicals' activities. So, maintaining iron at sufficient levels is very important to be monitored especially within the brain developments.

Keywords Brain development · Iron deficiency · Dietary iron sources · Iron homeostasis and/or dysregulations

Abbreviations

AD	Alzheimer's disease
CNS	Central nervous system
DCYTB	Duodenal cytochrome-b-like protein
DMT-1	Divalent metal transporter-1

N. A. Khalil (✉)
Nutrition and Food Sciences Department, Faculty of Home Economics, Menoufia University, Shibin El Kom, Egypt
e-mail: nazih.khalil@hec.menofia.edu.eg

W. Mohamed, R. Sandhir (eds.), *Trace Elements in Brain Health and Diseases*, Nutritional Neurosciences, https://doi.org/10.1007/978-981-99-1513-2_1

EFSA	European Food Safety Authority
FPN	Ferroportin
FSAI	Food Safety Authority of Ireland
Hb	Hemoglobin
IBR	Infant behavior record
ID	Iron deficiency
IDA	Iron deficiency anemia
IOM	Institute of Medicine
IRE	Iron regulatory element
IRE	Iron-responsive element
IRP	Iron regulatory protein
IRPs	Iron regulatory proteins
ROS	Reactive oxygen species
SACN	Scientific Advisory Committee on Nutrition
TBI	Transferrin-bound iron
TfR1	Transferrin receptor 1
TfRs	Transferrin receptors

1.1 Introduction

Trace elements known as trace metals are micronutrients mainly minerals needed in small or minor amounts for human health as they are present in all living tissues within all different ages. Most trace micronutrients elements are vital for the human body such as iron, manganese, selenium, zinc, etc. (Koller and Saleh 2018). They are divided nutritionally to be essential and nonessential elements while both of them required adequate and sufficient levels to meet physiological needs avoiding any malnutrition condition (Mehri 2020). They all are required crucially as cofactors with enzymes in order to catalyze in the cells different from many biochemical reactions and biological functions to ensure their proper functions. For instance, iron and copper in demand in energy metabolism for oxidation-reduction reactions and both play as catalyst in enzyme systems as primarily function. All such elements have only dietary sources (foods and nutrients) as they are not produced within the human body and their effects depending on human age, genetics, diseases, diet, lifestyle, and environment and these factors. Therefore, to optimize health status between different individuals, all elements should be achieved as recommended within dietary recommendations. Any nutritional deficiency with elements in total or in a single manner causes either impaired immunity or increase of infections' susceptibility as they all have therapeutic potential functions (Alanazi 2022). Indeed, previous study demonstrated that unbalanced or insufficient food mainly with trace element dietary intake levels either high or low could influence individual performance, poor healthy condition, thus low reproductive output. For instance, higher zinc and sodium concentrations in contrast to lower cesium and manganese concentrations show higher pregnancy and calf survival probabilities between female in addition to correlation between nutrition and individual performance that indicated trace element nutritional sources are important factors to be considered (Mozrzymas 2018). The nutritional deficiency in either infants and/or children will affect the nervous systems the most, while the impact in adults is mostly the brain which usually is spared the effects of malnutrition (Dhopeshwarkar 1983). It is indeed nutrition optimal brain of infant development and that is resistant rather to nutritional deficiencies as shown with psychomotor stimulation of the baby by iron and long-chain polyunsaturated fatty acids (Guesry 1998). Additionally, trace elements' role within the immunity such as zinc and selenium in addition to other trace elements illustrated an activation of the cytokine-mediated immune response that has been triggered TH1 cells synthesis with pro-inflammatory cytokines. Additionally, some other elements possess antiviral properties that interfering entry and replication of the viruses; thus, they show to have important role in immunity (Mozrzymas 2018). Also, iron plays an important role in oxygen transportation as a constituent of hemoglobin and myoglobin (National Research Council (US) Committee on Diet and Health 1989). Thus it is very important for many neurodegenerative process/developments that in total are all functional process needed for all human ages. Iron availability is varying due to many factors especially the human health state, infection and inflammatory stress in addition to nutritional and dietary consumption factors within different ages and genders under the iron homeostasis conditions. The current chapter aimed to point out the iron dysregulations with different relationship between iron status and neurological/brain developments or sensitive to iron-related neural changes. Also, neurophysiological function with cognitive and motor or socio-emotional development will be discussed within iron deficiency (ID) effects and different prevention strategies for protecting maternal health and brain developments especially at AD and PD conditions.

1.1.1 Iron as a Trace Element

Iron is very important trace element that is needed for not only all human genders and ages but also for all living organisms in order to maintain many different biological functions starting from not only the newborn babies but also the pregnancy stage. It is a transition metal that has the ability for transporting oxygen and transferring electrons. So, it plays important catalyst in cell growth and differentiation, oxidative, oxygenases, energy metabolism. Additionally, it is crucial for different many neurodevelopmental processes for maintaining neuronal tissues within the best metabolic and energetic requirements (Belaidi and Bush 2016). Thus, it is required for all cells with such different physiological processes; e.g., during the time of late fetal and early postnatal, it needed for hemoglobin (Hb) synthesis with different daily requirements depending on many factors. For instance, age and genders in addition to health state such as rapid growth and development period that is required high levels as many other nutrients as well. It is essential for cell functions

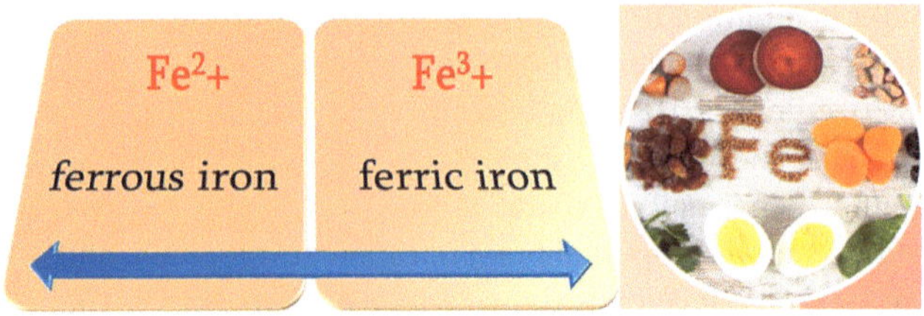

Fig. 1.1 Shows both iron forms and some photos of dietary sources; obtained and modified from centers for Disease Control and Prevention (2002)

mainly in brain developments such as myelination, energy, and neurotransmitter metabolism (Georgieff et al. 2018).

1.1.2 Dietary Iron Forms and Sources

Iron has different dietary sources that are mainly divided into two main sources; either with animal or plant foods/products. Both have different nutritional values with the best at the animal sources that are known by the hem sources and that would be well absorbed and used quickly within the healthy subjects/models (Peng et al. 2021). Most important for such trace element is the biological function that relies on its redox potential within the reversible transition; **ferrous (Fe^{2+})** to the **ferric (Fe^{3+})** state are only the two oxidation states that help in the reactions of catalyzing electron-transfer while the form of *ferrous (Fe2+) is the one required for iron* absorption (Ems et al. 2022) (Figs. 1.1 and 1.2).

Different studies shown that iron sufficiency plays an important role in pregnancy and infancy particularly for neurodevelopment processes especially infant ages at 0–6 months, 6–24 months, and/or children up to 4 years old (Wang et al. 2019; McCann et al. 2020). The pregnant women shown to have inadequate iron levels as daily allowances as in Europe that shown pregnant women of about 60–100% not meeting recommended intakes (Milman 2020). However, they have to enter the pregnancy stage with huge iron stores in addition to consuming high dietary iron sources with abundant in bioavailable iron in order to avoid the iron deficiency (ID; Lynch et al. 2018).

1.1.3 Iron Deficiency

Insufficient iron intake levels with pregnant women, babies at their first 1000 days, and young children show to have long-lasting deficits in cognition, motor function, and behavior because of their high requirements in order to support growth and development in addition to the low dietary consumed levels and lifestyle factors that

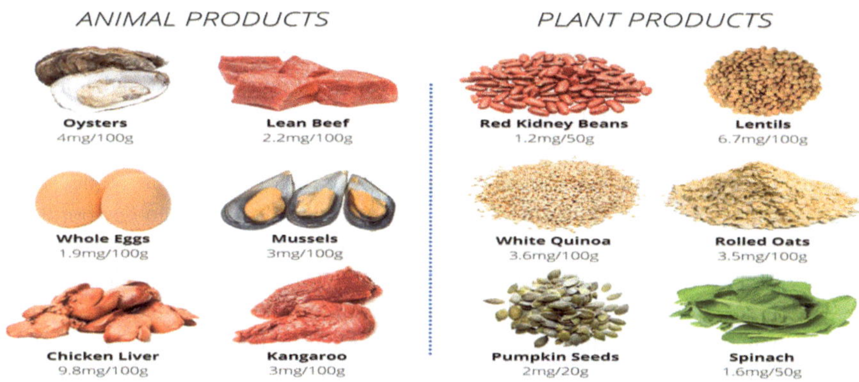

Fig. 1.2 Shows some different dietary iron sources (animal and plant samples). Data collected and modified from US Department of Agriculture, Agricultural Research Service (2019) & Tracey Frimpong (2021)

all in total resulting in ID (Lynch et al. 2018). The most important lifestyle factors are the dietary ones especially the role of different micronutrients in early brain development; mainly for the current chapter is the iron (Wang et al. 2019). Thus, the main objective of this chapter is to illustrate the dietary iron sources, supplementations, and daily requirement allowances affecting the neurological/brain developments (PocketPills 2020). It meant to investigate the iron-deficiency effects with different prevention strategies for protecting maternal health and brain developments particularly the iron-related neural changes; neurophysiological function with cognitive and motor or socio-emotional development (McCann et al. 2020). Finally argue for maintaining iron at sufficient levels is very important to be monitored especially within the brain developments.

1.1.4 Iron Requirements & Absorptions

The human body has iron in total ranged from 3 to 5 g that has stored in both liver and red blood cells equally about 1.8 g each. Also, different levels were found in macrophages, bone marrow, and muscles at 600 mg, 300 mg, and 300 mg respectively while very small levels have been recognized in both kidney and brain (Belaidi and Bush 2016). The two absorbable dietary iron forms (heme and nonheme) can be derived from dietary animal food sources such as meat, seafood, and poultry for heme iron that is most easily absorbed form (15–35%) while plants and iron-fortified foods as nonheme iron is the low absorbed form (10%; Ems et al. 2022). Dietary iron intake is absorbed of ferrous and heme iron within the normal metabolism into the human blood via the small intestinal in about 1–2 mg per/day. Afterward it will be transported to different parts of the human body in iron needs that could be regulated the iron metabolism, changes with ages in addition to genetic and environmental factors leading to different metabolic iron disorders (Peng et al. 2021). Starting from the breast-feeding infants within their early life stage, they can get nearly 0.35 mg/l iron from their human breast feeding in contrast to the neonate formula that only has 0.07 mg/kg every day that could lead to neonates' iron-deficiency anemia (IDA) especially if they are breastfed for more than 4–6 months. In such condition, neonates should be supplemented with either iron supplementations and/or iron-fortified complementary foods. Additionally, when newborns are diagnosed with low levels of iron status, they should be supplemented with high dietary iron levels (Wang et al. 2019). Indeed, preterm neonates with adequate iron storage at birth and up to 2 months old have been reported not to require any iron supplementations till such age then will face ID and its neurological consequence effects by different risk factors such as inadequate levels of dietary iron intake and/or poor absorptions within the early childhood especially with the high growth rate in such time. It also needed to be controlled for their effects on the brain development; brain has the highest growth rate during this period (the first 1000 days of life) affecting all immediate and later critical brain function (Georgieff et al. 2018). Therefore, recommended routine for iron supplementations with the children with IDA has been launched out by the WHO (Peng et al. 2021). Additionally, different authority organizations recommended the required iron levels for both pregnant women and infants and that are presented as following in Table 1.1 including; Food Safety Authority of Ireland (FSAI), Scientific Advisory Committee on Nutrition (SACN), European Food Safety Authority (EFSA), and Institute of Medicine (IOM; McCarthy et al. 2022).

Regarding the iron absorption, it is absorbed readily in both sides of the duodenum and upper jejunum. However, many factors are affecting such absorption by inhibiting or enhancing it. For example, the inhibitors include certain dietary compounds such as phytate (Phytic acid; phosphorus-containing compound) with their main dietary sources; plant-based diets (e.g., grains and vegetables with nuts and seeds) that show an effect on iron absorption (Centers for Disease Control and Prevention 2002). Also, drinks such as black and herbal tea, coffee, and wine in addition to different fruits and vegetables with legumes and cereals inhibited the

Table 1.1 Dietary Recommended dietary iron levels for both pregnant women and infants, reference values for iron (mg/d) during first 1000 days[a]

Categories	FSAI	SACN	EFSA	IOM
Women, 18 years	14	14.8	16	18
Pregnant women	15	14.8	16	27
Lactating	15	14.8	16	9
Infants, 0–3 months	1.7	1.7	–	0.27[b]
Infants, 4–6 months	4.3	4.3	–	0.27[b]
Infants, 7–12 months	7.8	7.8	11	11
Children, 1–3 years	8	6.9	7	7

[a]Dietary reference values presented as requirement daily allowances and values
[b]Mean an adequate intake; data obtained from McCarthy et al. (2022)

iron absorption because of their polyphenol levels. Again, dietary fiber sources are capable to decrease iron bioavailability (Adams et al. 2018).

Both phytates and polyphenols' inhibitors affect the absorption of nonheme iron only. However, heme and nonheme iron could be inhibited by different dietary factors such as presented calcium levels on the diets. Additionally, animal dietary protein sources that have been shown to inhibit iron absorption as well differ such as casein, whey, and egg whites with soy protein as an example of plant protein forms. While different dietary factors are also affecting iron inhabitation because of their main chemical composition such as spinach, chard, beans, and nuts because of their content of oxalic acid (oxalate) that acts to bind thus inhibits iron absorption (U.S. Department of Agriculture, Agricultural Research Service 2019). On the other hand, other dietary factors have been shown to enhance and increase the iron absorption levels such as special factor presented within meat, poultry, and fish consumptions. Again, vitamin C (ascorbic acid) that forms a chelate with ferric (Fe3+; nonheme) iron especially within low pH levels presented in the stomach and that remains soluble in the alkaline environment of the duodenum (Centers for Disease Control and Prevention 2002). To conclude up, many different dietary factors can influence iron absorption by decreasing or increasing it that in total will affect the iron homeostasis and that in turn will affect the brain development.

1.2 Iron Homeostasis and Brain Development

Brain development within all the neuronal processes needs many nutrients such as different vitamins and minerals (Khalil 2022). The age of two years shown the best time for brain development and any recognized dietary malnutrition presented long-term consequences of the mental health that reflected by the poorer academic achievements in addition to economic productivity low levels that in total associate with the poverty cycle. So, the low-income countries or families with malnutritional states will be at the highest levels of cognitive problems especially within the first days of life (McCann et al. 2020). One of the most important required nutrients that play an important role for brain development is iron especially with the neurotransmitter and energy metabolism; any deficiency will cause neurological adverse effects especially within the early life (McCarthy et al. 2022). Low dietary iron consumptions as in ID presented good relation to the cognitive outcomes not only as nutritional factors but also with many related factors such as poverty or social-economical and/or illness. Iron homeostasis is very effective processes at the cellular level for different cognitive outcomes involved in central nervous system (CNS) (McCann et al. 2020). When iron enters the brain, it is normally processed by the endocytosis through the blood–brain barrier. Also, as membrane-bound ceruloplasmin forms, ferroportin will be regulated and controlled the ferroxidase enzyme as the main copper-carrying protein in the blood (McCarthy et al. 2022). Iron especially the form of Fe^{3+} is the form that has been absorbed from food and then will be reverted to Fe^{2+} by duodenal cytochrome-b-like protein (DCYTB). Afterward it combines to be transported as ferrous iron into the intestinal epithelial cells and for heme molecule synthesis, or oxidized to ferric iron to mitochondria then be stored in ferritins. In the other hand, Fe^{2+} in plasma at the basal intestinal cell membrane released by ferroportin (FPN) and oxidated to Fe^{3+} again (Peng et al. 2021). Additionally, the iron (Fe^{3+}), greater part from transferrin in plasma is transported through blood that could be obtained under physiological conditions such as transferrin receptors (TfRs) that are expressed in neurons, then the neuronal uptake of transferrin-bound iron (TBI; TF-Fe complex) could be formed. Afterward it will be internalized; enters iron demanding cells by way of clathrin-mediated endocytosis (Wang et al. 2019; Peng et al. 2021) in addition to the way of TfRs commonly and by dicarboxylic acid receptor as well as by lactoferrin receptor. While neurons under inflammatory conditions with other glial cells and non-TBI from upregulated divalent metal transporter-1 (DMT-19) could be obtained. The metabolism of iron controlled within the cellular level posttranscriptionally by IRE (iron-responsive element)/IRP (iron regulatory protein; Wang et al. 2019). IRPs (IRP1 & IRP2) then will be bind for regulating the translation or stability of IRE-containing mRNAs encoded crucial iron metabolic proteins. Also, IRPs could be activated by ID and other stimuli to be blinded to cognate IREs, and inhibit specific translation of H-, L-ferritin, and ferroportin. Such condition with ID and/or IDA is associated well with poor neuronal/cognitive consequences especially within the early ages; newborns with many different motor and cognitive development such as maturation of the CNS, recognition memory, social-emotional behavior as following (Wang et al. 2019). Additionally, such condition is very common with ID in different brain disorders such as Alzheimer disease (AD).

Iron is stored and transported in protein-bound within physiological normal conditions that does not allow any free-radical reactions. On the other hand, any brain injury could cause protein-bound iron release inducing free-radical production especially when blood pH decreases. Such conditions will result in iron accumulation as seen within the injured neurons. Also, iron could be released within free radicals after mobilization from ferritin in addition to great quantity of nitric oxide production resulting from its binding protein in more iron (Wang et al. 2019). All such conditions in total will lead to extensive cellular oxidative stress and cell death resulting in different brain disorders.

1.2.1 Dietary Iron Supplementations and Brain Development

Iron as an important nutrient is needed for different psychological functions; cognitive ability, behavior, motor, socio-emotional, and neurodevelopments as it has a positive effect due to the enzymes containing iron, so consuming or supplementing sufficient dietary iron is very important positively especially within the early life spam (Wang et al. 2019).

1.2.1.1 Brain Development in Early Life

In early life for babies, all nutrients are needed and required for normal brain development in sufficient quantities and in adequate proportions. Such importance is started at the beginning of the prenatal period with different roles in psychological functions (Tierney and Nelson 3rd 2009).

Regarding the nutritional management within such early age (childhood; first few years of life) especially the dietary iron sources, it seems that any mother educational level affects not only the iron status of the child but also its psychomotor development. Overall, any malnutrition for both increases or decreases nutrients such as folic acid low intake and/or alcohol high intake can influence deleteriously brain development (Mehri 2020; Ems et al. 2022). General such development of the brain starts few weeks after conception that later on by early adulthood will be completed with the brain basic structure during the prenatal period followed by early childhood and finally it will continuously form neural networks and fully developed by time/age over the long term. However, the most important is during the early life; babies' nutritional status especially at iron deficiency will result in infection sensitivity and low muscular strength in addition to low ability to communicate (speech and face processing) with both the environment and social activities such as behavior contribution; mainly cognitive, motor and socio-emotional development as following (Tierney and Nelson 3rd 2009). Moreover, the brain development and school performance were associated to unbalanced trace element in either higher or lower levels presented within the early life to influence individual performance, poor healthy condition, and low reproductive output indicated the importance of nutritional factors to be considered.

1.2.2 Cognitive, Motor and Socio-Emotional Development

As it has been explained early that the brain needs different micronutrients such as iron and manganese that are mainly depending on the stage of the life cycle and the cell types that have been reported to be involved in CNS; decreasing levels of neurotransmitter by manganese accumulation thus influence the cognitive functions. Again, iron is needed for neurodevelopmental; neural processing speed (behavioral and cognitive) within cognitive and its consequences (Mehri 2020). Also, different minerals' dysregulation has been associated with brain diseases and neurodegeneration such as cognition loss and Alzheimer disease (AD), mainly copper toxicity and zinc deficiency.

At about 5 years old, early iron supplementation and neurocognitive development show a tendency with beneficial impact in preterm infants (Wang et al. 2019). On the other hand, low dietary iron intake or insufficient iron-rich foods show a significant risk factor for ID within 6–24-month-old infants especially within the early food introduction; unmodified cow's milk. Such insufficient dietary iron intakes have been illustrated between infants and young children in Ireland, the UK, and across Europe (McCarthy et al. 2022). Also, the iron supplementation and the cognitive performance relation has been assessed within many different studies and some of them shown improvements especially the participants at baseline IDA who have shown either solely and/or strong effects. Regarding the motor development, iron supplementation effects between infants aged 6–24 months old on motor development have been investigated within different studies and some of them reported an improvement within motor scores after iron supplementation. Furthermore, the infant behavior record (IBR) and behavior rating scale (BRS) that were used as behavioral component to measure the socio-emotional development after iron supplementations identified differences between supplemented groups in addition to any differences with IDA infants in contrast to non-IDA group (McCann et al. 2020). However, to prove this tendency, much more studies are needed in large cases.

1.3 Dietary Iron and Brain Disorders

1.3.1 Parkinson Disease

As it has been mentioned early that iron is mainly required for all brain roles and developments, brain iron homeostasis with either iron overload and iron deficiency in different aging-related neurodegenerative diseases such as Alzheimer's, Parkinson's. However, it also illustrated to have therapeutic effects within such diseases as it will be discussed with the following sections (Khalil 2022). Parkinson disease (PD) shown to have iron metabolism dysregulation and has been associated within oxidative stress and cellular damage affecting many different iron neurochemistry. For example, it has been shown to affect the normal physiological metabolic activities and functions such as mitochondrial respiration, synthesis of myelin, and neurotransmitter. PD well known as the second most neurodegenerative age-related disease reaches at the age of 65 by 1–2% of the population characterized by cognitive decline and dementia linked to the loss of dopaminergic neurons of aggregated α-synuclein protein mutations called Lewy bodies that have been presented within PD patients (Khalil 2022; Belaidi and Bush 2016). Iron shown α-synuclein conversion from α-helix to β-sheet with also accumulation in the brain of PD patients especially in neurons and glia of the SN and that has been associated within the PD severity. PD patients shown low ferritin levels in contrast to high iron loading within their SN. Such low levels of postmortem PD brains in the SN could be associated with sustained IRP1 activity caused by the lack of regulation. Additionally, both patients and animal models with PD show high intracellular iron levels correlated within serum iron (Belaidi and Bush 2016). Such effect could be treated and/or controlled by the iron chelation as we will discuss later for both PD and AD.

1.3.2 Alzheimer Disease

The dementia common cause is AD that is characterized as a neurodegenerative age-related disease (incidence rate increasing by increasing the human life span) mainly by different brain functions such as impaired cognitive function, ability of learning reductions in addition to memory and motor loss that in total show strange behavioral symptoms (Khalil 2022). Such symptoms in total of dementia are characterized with memory and language difficulties with especially association of AD pathogenesis and imbalances of different trace elements such as aluminum (Al), copper (Cu), zinc (Zn), and iron (Fe). Also, some researchers suggested that the main AD causes could be iron with high redox activities that is related to the deposition of amyloid plaques in addition to the formation of nerve fiber tangles. Such iron metabolism disorder will catalyze reactive oxygen species (ROS) formation in the body in addition to more chemical reactions such as DNA attack with protein and lipid molecules leading to cell damage, oxidative stress in the brain, and AD formation (Peng et al. 2021). Indeed, the pathological process of AD shows correlation to oxidative stress. It has been induced by the disorder of iron metabolism leading to neuronal damage. Iron ions tend to be accumulated in the brain by increasing the human ages inducing oxidative stress additionally to some key iron homeostasis regulators. For example, ferritin protein and transferrin protein that are causing damage of neurons then resulting a decline in cognitive and memory functions (Peng et al. 2021). Also, iron-bound melanin transfer protein presented high levels in AD serum indicating that iron in the brain of AD patients might be bind abnormally. In contrast, different studies found regional deposition of iron in AD patients' brain and has been treated by iron chelator that alleviates AD symptoms effectively (Cassidy et al. 2020). Additionally, copper that shows a vital role because of being effective within many enzymes such as cytochrome c and Cu/Zn superoxide dismutase and any dyshomeostasis can cause increased oxidative stress levels with neuronal loss. Also, in order to carry out the CNS-required functions, the trace elements must be in balance in both body and brain tissues and any lack will be involved in the brain pathophysiology diseases, e.g., PD, AD with mental and cognitive disorders. Finally, such collected data supporting that there is a close relationship between iron metabolism disorder and AD. However, the main causes and pathogenesis have not been fully clarified, thus different therapeutic approaches and/or strategies supporting iron usage supplementation to treat AD conditions are as discussed as following.

1.4 Iron for AD and PD Treatment

1.4.1 Iron Chelation

Iron chelation is a used method to limit and redistributing the iron within the CNS directly using chelating agents such as deferoxamine, deferiprone, and ferrite. However, such chelating agents can improve the AD symptoms they can have toxic effects as in allergic reactions, liver and kidney failure, etc. (Peng et al. 2021). Reducing iron levels methods in addition to toxicity prevention is the main issue by using such iron chelation treatment efficacy for several neurodegenerative disorders such as PD conditions. Iron overload as well treated by deferiprone that affects the levels of iron positively especially at long-term treatment (Belaidi and Bush 2016). Indeed, PD patients at early-stage are treated with deferiprone in randomized, placebo-controlled clinical trial (30 mg/kg/day for a year shown motor handicap progression reduction in addition to iron deposits reductions). PD and AD have been associated well with iron especially the impaired iron exportation; either iron accumulation and neuronal loss in SN or cognitive deficits, and parkinsonism that could be prevented by iron chelation (Belaidi and Bush 2016).

1.4.2 Antioxidant Therapy

The oxidative stress induction within brain disorders especially AD by the excess iron metabolism leads to brain damage, neurotoxicity (DNA, protein, and lipids). It is also correlated well to the ROS scavenging, so using different antioxidants could be more helpful for getting rid of free radicals in the matrix. Many dietary antioxidant sources such as tea polyphenols show an effective AD improvement via their chelating iron, scavenging oxygen-free radicals, and anti-inflammatory properties and/or activities (Peng et al. 2021). Indeed, tea includes catechins that shown potential preventive effects for AD symptoms by its anti-inflammatory and antioxidant properties in both in vivo and in vitro (Ide et al. 2018). Different more dietary antioxidant compounds such as curcumin and caffeine (Peng et al. 2021) show more effects, however, more human studies are needed with different dietary antioxidant sources.

1.5 Conclusion

Nutrients are in early life as particularly important period of brain development especially dietary iron sources. Iron is very effective trace element needed for brain development especially within different brain disorders such as AD and PD. Additionally iron deficiency with risk of iron overload is correlated well with different brain function as iron is needed for CNS. However, excess iron is a potent source of oxidative damage; so, studying iron metabolism within the human brain could find new effective therapeutic strategies to improve different brain diseases. The iron metabolic alterations in the brain affected the neuronal functions as the dysregulation in Alzheimer′s and Parkinson′s diseases. Also, distribution of ID prevalence and importance of its treatment are very helpful for brain-metabolism recovery in addition to the role of iron especially at anxiety and depression treatments suggesting therapeutic roles in the etiology of brain disorders such as AD and PD. However, much more human studies regarding dietary factors providing essential micronutrients; iron and antioxidant supplementations within different brain disorders are needed within the use of antioxidant dietary foods or functional food.

References

Adams S, Sello C, Qin G-X, Che D, Han R (2018) Does dietary fiber affect the levels of nutritional components after feed formulation? Fibers 6. https://doi.org/10.3390/fib6020029

Alanazi S (2022) Trace elements and immunity: a brief overview. Res J Biotechnol 17(8):144–150. https://doi.org/10.25303/1708rjbt1440150

Belaidi AA, Bush AI (2016) Iron neurochemistry in Alzheimer's disease and Parkinson's disease: targets for therapeutics. J Neurochem 139:179–197

Cassidy L, Fernandez F, Johnson JB, Naiker M, Owoola AG, Broszczak DA (2020) Oxidative stress in Alzheimer's disease: a review on emergent natural polyphenolic therapeutics. Complement Ther Med 49:102294

Centers for Disease Control and Prevention (2002) Iron deficiency—United States, 1999-2000. Morb Mortal Wkly Rep 51:897–899

Dhopeshwarkar GA (1983) Effects of malnutrition on brain development. In: Nutrition and brain development. Springer, Boston, MA. https://doi.org/10.1007/978-1-4684-4280-9_7

Ems T, St Lucia K, Huecker MR (2022) Biochemistry, iron absorption. StatPearls Publishing. https://www.ncbi.nlm.nih.gov/books/NBK448204/

Frimpong T (2021) Iron nutrition summary for dietitians. Nutrients

Georgieff MK, Ramel SE, Cusick SE (2018) Nutritional influences on brain development. Acta Paediatr 107:1310–1321

Guesry P (1998) The role of nutrition in brain development. Prev Med 27(2):189–194., ISSN 0091-7435. https://doi.org/10.1006/pmed.1998.0292

Ide K, Matsuoka N, Yamada H, Furushima D, Kawakami K (2018) Effects of tea catechins on Alzheimer's disease: recent updates and perspectives. Molecules 23:2357

Khalil AN (2022) Malnutrition of micronutrients and brain disorders. Springer, pp 167–182

Koller M, Saleh HM (2018) An introduction to trace elements, trace elements-human health and environment. IntechOpen

Lynch S, Pfeiffer CM, Georgieff MK et al (2018) Biomarkers of nutrition for development (BOND)-iron review. J Nutr 148:1001s–1067s

McCann S, Perapoch Amadó M, Moore SE (2020) The role of iron in brain development: a systematic review. Nutrients 12(7):2001. https://doi.org/10.3390/nu12072001

McCarthy E, Murray D, Kiely M (2022) Iron deficiency during the first 1000 days of life: are we doing enough to protect the developing brain? Proc Nutr Soc 81(1):108–118. https://doi.org/10.1017/S0029665121002858

Mehri A (2020) Trace elements in human nutrition (II)—an update. Int J Prev Med 11:2. https://doi.org/10.4103/ijpvm.IJPVM_48_19

Milman NT (2020) Dietary iron intake in pregnant women in Europe: a review of 24 studies from 14 countries in the period 1991–2014. J Nutr Metab 2020:7102190
Mozrzymas R (2018) Trace elements in human health. In: Chojnacka K, Saeid A (eds) Recent advances in trace elements. https://doi.org/10.1002/9781119133780.ch18
National Research Council (US) Committee on Diet and Health (1989) Diet and health: implications for reducing chronic disease risk. National Academies Press (US), Washington (DC). 14, Trace Elements. https://www.ncbi.nlm.nih.gov/books/NBK218751/
Peng Y, Chang X, Lang M (2021) Iron homeostasis disorder and Alzheimer's disease. Int J Mol Sci 22(22):12442. https://doi.org/10.3390/ijms222212442
PocketPills (2020) Birth control pills and iron deficiency. PocketPills. https://www.pocketpills.com/blog/birth-control/birth-control-pills-and-iron-deficiency
Tierney AL, Nelson CA 3rd (2009) Brain development and the role of experience in the early years. Zero Three 30(2):9–13. PMID: 23894221; PMCID: PMC3722610
U.S. Department of Agriculture, Agricultural Research Service (2019). Food Data Central. fdc.nal.usda.gov
Wang Y, Wu Y, Li T, Wang X, Zhu C (2019) Iron metabolism and brain development in premature infants. Front Physiol 10:463. https://doi.org/10.3389/fphys.2019.00463

Chapter 2
Trace Elements and Mild Cognitive Impairment

Ke Tian Yong and Shi-Hui Cheng

Abstract As an intermediate stage between normal age-related cognitive decline and dementia, mild cognitive impairment (MCI) serves as an ideal time interval for the treatment that can potentially arrest or even reverse the cognitive symptoms. Trace elements are critical for brain function, and they partake in redox reactions and metabolic processes in the central nervous system (CNS). While a small amount of them is required to protect the neuronal tissues, dysregulation of trace element homeostasis may lead to the development of pathologic states. Overexposure to various trace elements has been postulated to play a role in the onset and progression of neurodegenerative diseases, thus understanding how these metals influence the metabolic pathways that can be useful in the prevention of MCI or its progression to dementia. This chapter presents the pathology of MCI and the factors that have been shown to link to the development of MCI to explore the relationship between trace elements—calcium, copper, zinc and selenium—and neurodegeneration. It is evident that both deficiency and overexposure to these metals cause oxidative stress and that oxidative damage is the most explored mechanism in its association with neuronal function. Dysregulated homeostasis of these metals is also linked to Aβ and/or tau aggregation, which may further worsen the condition, although some protective effects were observed via the actions of zinc and selenium. While supplementation has shown mixed results, potential interventions to delay cognitive decline can be formulated via further research on the role of trace elements in CNS function and neurodegenerative pathology.

Keywords Trace element · Mild cognitive impairment · Calcium · Copper · Zinc · Selenium · Neurodegeneration · Cognitive decline · Nutrition

---

K. Yong · S.-H. Cheng (✉)
School of Biosciences, Faculty of Science and Engineering, University of Nottingham Malaysia, Semenyih, Selangor, Malaysia
e-mail: ShiHui.Cheng@nottingham.edu.my

W. Mohamed, R. Sandhir (eds.), *Trace Elements in Brain Health and Diseases*, Nutritional Neurosciences, https://doi.org/10.1007/978-981-99-1513-2_2

2.1 Introduction—Ageing, Cognitive Decline and Nutrition

Owing to improved healthcare and technological advances, human life expectancy has shown an unprecedented increase, which simultaneously makes population ageing an undeniable reality (Power et al. 2019). According to the Baseline Report for the Decade of Healthy Ageing, people aged 60 years and above took up 13.5% of the total population (more than 1 billion) in 2020 and the number is expected to project to 2.1 billion (equivalent to 1 in 5 people) by 2050 (World Health Organization 2020).

As the longevity of the population increases, healthy ageing becomes essential to sustaining a good quality of life. While functional ability has to be optimised to ensure well-being in old age, the amount of people with cognitive decline continues to rise exponentially alongside the ageing population, posing an enormous challenge to fostering the healthy ageing process (Dominguez and Barbagallo 2018; World Health Organization 2020). Meanwhile, this increasing prevalence of age-related neurodegenerative disorders also indicates a greater social and economic burden as dementia is the major contributor to disability among older adults (Dominguez and Barbagallo 2018; Ogawa 2014).

Understanding the biology behind the ageing process and the association between ageing and cognitive decline is critical to contriving strategies for healthy ageing. Modulated by numerous genetic and molecular mechanisms, ageing is a complex and progressive process that leads to functional and structural changes in an organism (Bhatti et al. 2020; Power et al. 2019). With age, body cells that can once replicate themselves now fail to divide as they have achieved the Hayflick limit. As more cells enter programmed cell death and various molecular and cellular damages accumulate, it lowers the individual's physical and mental capacity, at the same time elevates the risk of diseases (Power et al. 2019; World Health Organisation 2021). This replicative capacity is thought to be limited to cells in the brain, in which the rate of neurogenesis drops as a person ages and is only restricted to certain brain regions in adulthood (Catlow et al. 2016; Power et al. 2019). Parenthetically, glial cells and neurons are considered to be more vulnerable to dysfunction and structural loss, especially when they are also sensitive to free radicals (Power et al. 2019; Uttara et al. 2009).

Despite being a major oxygen metaboliser (taking up approximately 20% of the body's oxygen intake), the brain is equipped with a relatively poor antioxidant defence system and is susceptible to attacks from the reactive oxygen species (ROS) (Popa-Wagner et al. 2013; Ogawa 2014). Moreover, ROS is particularly active in the neuronal tissues, acting as a source of oxidative stress that causes neuronal damage, which serves as the fundamental basis for neurodegenerative disorders (Uttara et al. 2009; Ogawa 2014). Other than the build-up of oxidatively damaged molecules, mitochondrial dysfunction, impaired DNA repair and inflammation are also some of the hallmarks of brain ageing (Mattson and Arumugam 2018).

As ageing progresses, the brain undergoes some physical and cognitive changes, in which the combination of these structural and functional factors is likely to result in the degradation of neurocognitive function (Broglio et al. 2012). One of the structural changes that occur will be cerebral atrophy, which refers to a reduction in brain volume. This condition does not develop equally for all brain areas and the effects are most prominently shown in the prefrontal cortex and in certain studies, the hippocampus (Murman 2015; Peters 2006). A drop in neuronal volume has been proposed to be the causative factor of alterations in the brain rather than the number of neurons (Peters 2006).

Besides, structural changes such as loss of synapses, reduced length and number of dendrites, and decreased amount of axons have been observed along with ageing (Murman 2015). Specifically, the white matter that consists of myelinated nerve fibres and is responsible for transmitting nerve signals within the brain appears to have a more pronounced reduction in volume and integrity compared to the grey matter (Power et al. 2019). Other morphology changes include cortical thinning, ventricular enlargement and loss of gyrification (Blinkouskaya et al. 2021).

Cognitive abilities are important for functional independence and tend to deteriorate with age as the neuronal network changes (Murman 2015). Cognition encompasses mental processes involving knowledge acquisition, retention of information and integration of such processes into responses. These include memory, learning, perception and decision-making (Dauncey 2014; Kihlstrom 2018). Deterioration in prospective memory (memory for intended actions), complex attention task performance and executive cognitive functions such as multitasking, inhibiting and decision-making are common cognitive changes linked to ageing (Murman 2015). Meanwhile, several cognitive domains such as language, speech and implicit memory (long-term memory for skill learning) are comparatively stable throughout ageing (Murman 2015; Power et al. 2019).

There is a huge variation in clinically meaningful cognitive decline between individuals. This clinical variability is said to be depending upon one's cognitive reserve, in which individuals with more neuronal connections, denser grey matter, and a stronger neuronal network are able to cope with the damage and avert symptoms of neurodegenerative changes to continue functions like usual (Broglio et al. 2012; Stern 2012). Thus, different consequences can result from the same amount of brain damage for different people, and the rate of age-associated cognitive decline also varies among individuals, influenced by multiple behavioural and environmental factors (Broglio et al. 2012).

As brain pathology takes up years or decades to develop before the cognitive symptoms appear, applying the right strategies to delay the onset of neurodegenerative disorders is essential (Dominguez and Barbagallo 2018). Nutrition is recognised as one of the potential preventive measures for age-associated cognitive decline, in which an association has been suggested between one's nutritional status and cognitive impairment (Ogawa 2014). A number of reviews have discussed the roles of specific nutrients linked to cognitive ageing or neurocognitive development, including B vitamins (folate and cobalamin), polyunsaturated fatty acids (PUFA) and antioxidants (vitamin C, tocopherols, carotenoids, phenolic compounds and minerals required for antioxidant enzymes) (Del Parigi et al. 2006; Dominguez and Barbagallo 2018; Tucker 2016). A recent review on natural food products such as edible bird's nest revealed its neuroprotective properties, with studies reporting its ability to suppress neuroinflammation and neuronal cell death due to its sialic acid content (Loh et al. 2022).

Being the predominant lipid present in the metabolically active parts of the brain, docosahexaenoic acid (DHA) is vital for the neuronal cell membranes for its fluidity, enabling neurotransmitter communication (Dominguez and Barbagallo 2018). Its favourable effects are often discussed with the consumption of fish, in which their cardiovascular protective effects may play a great role in the prevention of vascular dementia (Morris 2012). On the other hand, homocysteine accumulation has been associated with cognitive decline. The studies of folate and cobalamin deficiency have been of interest, mainly on their roles as cofactors in homocysteine metabolism. The supplements of folic acid and cobalamin have been reported to be effective in reducing plasma homocysteine levels (Morris 2012; Tucker 2016).

In addition, the synergistic interactions that exist in various dietary patterns were postulated to be responsible for their benefits on cognitive functions (Dominguez and Barbagallo 2018; Tucker 2016). Nutrients may provide a greater performance when acting as a part of a food matrix or dietary pattern rather than existing alone, which explains the inconsistent outcomes for isolated nutrients in clinical trials (Tucker 2016). Thus, many studies have been shifting the focus from examining individual nutrients to the whole diet for neuroprotective functions (Tucker 2016). Adherence to neuroprotective dietary patterns such as the Mediterranean diet and Dietary Approach to Stop Hypertension (DASH) were associated with a reduced rate of cognitive decline and incident Alzheimer's disease (AD) (Dominguez and Barbagallo 2018). The Mediterranean diet is the most studied dietary pattern in this case, and it is suggested to exert various benefits including the preservation of structural brain integrity (Tucker 2016).

As previously mentioned, oxidative stress in the brain is comprehensive as it plays a major role in cognitive deterioration. Micronutrient deficiency or excess is one of the factors that can lead to genome damage. Micronutrients are vital for normal body function, acting as enzyme cofactors or a component of metalloenzymes, which are involved in DNA replication, DNA methylation and prevention of DNA damage (Meramat et al. 2015). As elements like iron, zinc, selenium and copper are responsible for maintaining antioxidant performance and affecting genes that are involved in protein coding, a lack of these nutrients in the diet can bring a change in biological functions, and this deficiency has also been linked to cognitive decline (Lo and Cheng 2023) Conversely, metals in excess can deposit in the brain and cause oxidative injury. Metal ion homeostasis is vital for neuronal function and changes in the balances of these ions have been proven a risk factor for neurodegenerative diseases (Grochowski et al. 2019; Purushothaman et al. 2020).

Therefore, this chapter aims to discuss the current findings on age-related cognitive decline, specifically mild cognitive impairment (MCI), and its relationship with nutrition, with highlights on the role of several trace elements including calcium, copper, zinc and selenium. To better understand the link between cognitive function and these micronutrients, the factors that are involved in the pathophysiology of

MCI including oxidative stress, DNA damage, mitochondrial dysfunction and telomere shortening are also addressed in this chapter.

2.2 Mild Cognitive Impairment (MCI)

2.2.1 Prevalence of MCI

Cognitive decline can be divided into different degrees of severity, with MCI conceptualised as the transitional stage between normal ageing and dementia (Morris et al. 2001). Memory deficit is a common complaint for the affected individuals, with the absence of significant impairment in any instrumental activity of daily living. In the other words, MCI is a state where the manifested cognitive impairments surpass the expected age-related decline in cognition, yet are not enough to meet the criteria for diagnosis of dementia (Cabral Pinto et al. 2019; Del Parigi et al. 2006; Sanford 2017). MCI involves a slight but detectable decline in one or more cognitive domains, usually referring to the ability to recall information or learn and retain new information (López-Sanz et al. 2017; Sanford 2017). In some studies, subjective cognitive decline (SCD) is also recognised as a preclinical asymptomatic stage of AD prior to MCI, but this remains a debate (López-Sanz et al. 2017).

Over the years, risk factors of MCI have been established via multiple epidemiologic studies, including old age, genetics, low educational level, physically or cognitively sedentary lifestyle, male sex, presence of apolipoprotein E4 (APOE-4) allele and vascular risk factors such as coronary artery disease, hyperlipidaemia, stroke and hypertension (Jongsiriyanyong and Limpawattana 2018; Ritchie 2004; Sanford 2017). Besides, the risk of MCI has been shown to be associated with the number of concurrent conditions, in which older adults with multimorbidity have 1.38 times the risk of MCI compared to those with one or none (Vassilaki et al. 2015). It is critical to recognise the practical interventions to reduce exposure to modifiable risk factors, especially when the existing conditions have been worsened by COVID-19, which has raised AD and dementia deaths by 17% in 2020 (Alzheimer's Association 2022a).

To formulate recommendations on the diagnosis and treatment of MCI, the prevalence of MCI within the general population will be one of the main clinical questions posed (Petersen et al. 2018). The prevalence of MCI, however, differs greatly across different studies. While a review revealed the range as 12% to 18% in individuals above 60 years based on several international studies (Petersen 2016), another review reported a prevalence as low as 3% to 22% for those over 65 years, also based on studies investigating different demographics of population (Sanford 2017). Ward et al. (2012) have even found a greater range of prevalence, which is from 3% to 42% across 35 publications inclusive of systematic reviews and population-based observational studies.

In fact, it is challenging to determine the exact prevalence of MCI from studies due to the different definitions of MCI used, methodological variability and unstandardised diagnostic criteria implemented (Petersen 2016; Sanford 2017). There are variations in the breadth of MCI considered (amnestic, non-amnestic or all MCI), data collection methods, geographical region, sample size, population selected and length of follow-up, which explains the differences among the reported estimates (Petersen 2016; Ward et al. 2012). Most importantly, the operational definitions of the cognitive impairment subtypes should be enhanced, as this allows a better understanding of the disease burden and identification of people at risk of developing future dementia (Ward et al. 2012).

Although MCI is an at-risk stage of AD, MCI is not synonymous with AD and not all cases progress to dementia (Sanford 2017). According to the Alzheimer's Association (2022a), around one-third of individuals having MCI due to AD will progress to dementia five years after diagnosis. Meanwhile, the annual progression rate of MCI to dementia was reported as 5% to 17% by Jongsiriyanyong and Limpawattana (2018). Another review recorded a conversion rate of 10% each year, and it elevates to 80–90% after 6 years (Ataollahi Eshkoor et al. 2015). The good news is that many individuals are likely to revert to normal status or show no progression to dementia, in which studies have shown a 30–50% conversion rate for MCI patients reverting to normal in the next follow-up (Ward et al. 2012).

2.2.2 Pathogenesis

MCI is neuropathologically complex. While the key drivers of neurodegeneration and the events secondary to the disease progression are still yet to be well determined, studies have reported a number of pathological changes in MCI (Stephan et al. 2012). These events include disrupted protein metabolism, synaptic dysfunction, mitochondrial changes, formation of plaques and tangles, neurochemical deficits and cellular injuries (Stephan et al. 2012).

Neuropathological studies have suggested that the interaction between tau tangles and amyloid-β (Aβ) plaques, along with other factors, were responsible for the development of AD related brain changes (Alzheimer's Association 2021; Small et al. 2006; Stephan et al. 2012). The soluble building blocks of Aβ and tau are capable of self-propagation by prion-like mechanisms (Bloom 2014). During ageing, it was observed that these microscopic brain protein fragments greatly build up and the prevalence of the lesion formation elevates as the individual grows older (Alzheimer's Association 2021; Small et al. 2006).

Tau tangles are also known as neurofibrillary tangles (NFTs). These abnormal aggregates are formed when tau proteins detached from microtubules and stick to other tau proteins inside neurons (National Institute on Aging 2017). This compromises the structural integrity of the internal skeleton as tau proteins are essential for microtubule stability. By blocking the neuron's transport system, nutrients and other substances fail to travel in the neurons, which impair the synaptic transmission between neurons (Alzheimer's Association 2021; National Institute on Aging 2017). Mechanisms involved in the formation of neurofibrillary lesions are possibly related to protein phosphorylation, as studies found a large proportion of hyperphosphorylated microtubule-associated protein tau as the main constituent of NFTs (Brion et al. 2001; Metaxas and Kempf 2016).

NFTs have been discovered in patients with MCI, mainly in the hippocampus and certain medial temporal regions, and will continue to spread to other areas of the brain as it progresses to AD (Small et al. 2006). Analysing the cerebrospinal fluid (CSF) is useful to provide information on the brain changes resulting from tauopathy (Alzheimer's Association 2021). Studies have found a correlation between the tau markers in CSF and hippocampal atrophy, a feature that strongly predicts future cognitive decline (Chiu et al. 2014; Goukasian et al. 2019). Additionally, a negative association has also been found between plasma tau concentration and verbal fluency, memory performance, and visual reproduction, indicating the presence of neuronal damage (Chiu et al. 2014; Dage et al. 2016).

Simultaneously, people with MCI will have neuritic plaques (Aβ deposits) accumulating within the neocortical regions and hippocampus, but their role as a primary contributing factor remains an argument (Chiu et al. 2014; Small et al. 2006; Stephan et al. 2012). Aβ peptides are derived from amyloid precursor protein (APP), a transmembrane protein that takes part in activities such as signalling and neuronal development, via the action of β-secretase and γ-secretase (Chen et al. 2017). Extracellular senile plaques have Aβ as their predominant protein component, in which the peptides assemble to form the insoluble amyloid fibrils, and then further accumulate into plaques (Chen et al. 2017).

While an increase of tau in CSF is indicative of AD, low levels of Aβ42 in CSF also serve as a biomarker, which denotes the entrapment of Aβ within plaques (Chiu et al. 2014). In the CSF, major forms of Aβ discovered are Aβ38, Aβ40 and Aβ42 (Mawuenyega et al. 2013). Among all, Aβ42 is the most studied isoform, which is also believed to be more neurotoxic and appears to play a more important role in plaque formation (Jäkel et al. 2019; Mawuenyega et al. 2013). As the species that dominates in senile plaques, Aβ42 has a high hydrophobicity and is more aggregation-prone compared to Aβ40, the isoform that exists in a much greater abundance in the brain (Chen et al. 2017; Gu and Guo 2013).

The Amyloid cascade hypothesis is a model that helps explain the pathogenesis of AD (Sturchio et al. 2021). This hypothesis postulates that Aβ production and aggregation are the initial events that trigger a cascade of neurotoxic events, such as neuroinflammation and intracellular tau accumulation, eventually leading to cognitive impairment (Sturchio et al. 2021). It was suggested that an imbalance between Aβ generation and degradation is the origin, in which more normal proteins and peptides are misfolded into insoluble fibres, causing loss of function and cell death (Mawuenyega et al. 2013; Sturchio et al. 2021).

As a genetic risk factor for MCI, APOE-4 has also been found to be associated with Aβ and tau pathology. Benson et al. (2022) have observed higher levels of total tau and phospho-tau, as well as lower levels of CSF Aβ42 in APOE-4 carriers compared to the non-carriers, and that the effects of APOE-4 were found only in

MCI-stable subjects. Nonetheless, the link between APOE-4 and Aβ and tau accumulation may not be the full picture (American Association for Clinical Chemistry 2020). There is growing evidence that supports the role of APOE-4 in blood–brain barrier (BBB) dysfunction, in which it causes destruction to the brain capillary pericytes and speeds up the breakdown of BBB (Montagne et al. 2020). According to a study conducted by Montagne et al. (2020), the presence of BBB breakdown within the medial temporal lobe and hippocampus was found in APOE4 carriers and not those without APOE4, with an even higher severity for subjects with cognitive impairment.

2.2.3 Symptoms and Treatment

Symptoms of MCI can be classified based on the type of MCI-amnestic (aMCI) and non-amnestic MCI (naMCI), which differs by the thinking skills or cognitive domains affected (Alzheimer's Association 2022b). Individuals with aMCI have significant memory disturbance. aMCI is associated with a greater risk of progression to AD, and it shares the same memory loss feature with dementia, but with less severe functional impairment (Csukly et al. 2016; Ellison et al. 2008). The patient may forget recent events or important information, which would have been easily recalled earlier, such as conversations and appointments (Alzheimer's Association 2022b).

On the other hand, naMCI involves impairment of cognitive abilities other than memory, including executive functions such as organisation and planning, attention and visual depth perception (Dugger et al. 2015). Therefore, the common symptoms for people with naMCI will be having trouble looking for the right words, being easily distracted and struggling to judge distances and interpret 3D objects (Alzheimer's Society 2022). In addition, individuals with MCI may also experience neuropsychiatric symptoms (NPS). The symptoms tend to take place simultaneously, and the effects from some combinations can be cumulative on subsequent cognitive decline. The most common NPS have been summarised in reviews, which include sleep problems, anxiety, depression, apathy and irritability (Martin and Velayudhan 2020; Monastero et al. 2009).

Dementia has a potentially large treatment window, in which subtle changes in cognitive function may already exist 20 years before disease onset (Ritchie 2004). Thus, treatment at MCI, the transitional state between normal ageing and AD, has a particular significance in delaying or preventing the potential development of dementia (Morris et al. 2001). Moreover, many cases of MCI are non-progressive and will revert to normal cognition, which also explains why MCI became a target for clinical trials to uncover new therapies (Sanford 2017; Cabral Pinto et al. 2019).

Although MCI is seen as a point of intervention and pharmacological agents such as cholinesterase inhibitor drugs are being tested in MCI patients in multicentre trials, none of the conducted trials demonstrated effectiveness and the Food and

Drug Administration (FDA) has not accepted any medication for MCI treatment (Morris et al. 2001; Petersen 2016). Instead, the research on non-pharmacological therapies, particularly lifestyle modifications, has shown some progress (Petersen 2016). The recent focus of intervention includes exercise, cognitive stimulation and dietary changes, which rely greatly on patients' motivation (Sanford 2017).

It has been suggested that physical activity like aerobic exercise is useful in reducing the rate of conversion from MCI to dementia (Petersen 2016; Vega and Newhouse 2014). A systematic review supported this protective effect, with a highlight on the pronounced effect of moderate-intensity aerobic exercise on global cognition, and the effectiveness of aerobic exercise in improving working memory (Law et al. 2020). Another review that included a total of 10 randomised controlled trials from East Asia, Europe and South Asia also reported a significant improvement in cognitive performance in older adults with MCI upon engaging in regular aerobic exercise, but with limitations such as the inclusion of diverse types of aerobic exercise (Yong et al. 2021). While the exact mechanism is undetermined, the neuronal health benefits from exercising were suggested to result from the stimulation of nutrients and oxygen delivery to the brain, the release of neurotrophins and its stress relief effects (Sanford 2017). Similarly, cognitive training may benefit MCI patients by exerting a neuroprotective effect. Jigsaw puzzles, card games and word search puzzles are some of the examples of activities that cause mental stimulation (Sanford 2017).

In the aspect of diet modification, the Mediterranean diet (MedDiet) has shown promising results in lowering the incidence of cognitive decline (Sanford 2017). Consistent adherence to the MedDiet has been shown to slow MCI progression, and the relationship between individual components of the diet, such as fatty acids and fish, and MCI has also been explored (Sanford 2017; Scarmeas et al. 2009). For example, fish oil rich in DHA and eicosapentaenoic acid (EPA) was found to improve depressive symptoms after six months of intervention, which implies its potential use to boost mood symptoms in MCI patients (Vega and Newhouse 2014). Nevertheless, there are pharmacological treatment studies, which have failed to prove any effectiveness of omega-3 fatty acids as a treatment for MCI (Vega and Newhouse 2014). Besides, supplementation of vitamins B and E and ginkgo biloba appears to exert no significant effect on MCI progression, except when a clear vitamin deficiency exists (McGrattan et al. 2018; Sanford 2017; Vega and Newhouse 2014). For all the non-pharmacological factors discussed, it is important to note that conflicting results still exist and further exploration is necessary (Demurtas et al. 2020; Lopez 2013; McGrattan et al. 2018; Vega and Newhouse 2014). Well-designed large-scale studies with bigger sample sizes will be useful to ascertain their effects and clinical importance.

2.3 Factors Associated with MCI

2.3.1 Oxidative Stress and MCI

Cellular redox homeostasis is of utmost importance for normal cell function (Gambhir et al. 2022). It is a dynamic process that works to maintain the balance between the magnitude and rate of oxidants produced and their elimination within the cell (Gambhir et al. 2022; Le Gal et al. 2021). Hydrogen peroxide (H_2O_2), nitric oxide (NO), hydroxyl radical (OH•) and superoxide (O_2^-) are some of the well-known oxidants. The cellular redox balance can be disrupted due to excessive generation of oxidants as well as a reduction in antioxidants or impairment of the antioxidant defence systems (Swomley and Butterfield 2015). These antioxidant cellular defences can be enzymatic or non-enzymatic, which includes the reduced nicotinamide adenine dinucleotide phosphate (NADPH)-regenerating systems, thioredoxin, glutathione and its associated enzymes such as glutathione reductase, glutathione peroxidase and glutathione-S-transferase (Le Gal et al. 2021; Swomley and Butterfield 2015). For example, copper–zinc superoxide dismutase (Cu/Zn-SOD) is a system that scavenges O_2^- by converting it to H_2O_2 to prevent further damage (López et al. 2013).

Redox interactions are involved in various biological processes; these include immune responses, cell differentiation and metabolism (Gambhir et al. 2022; Le Gal et al. 2021). A dysregulation of oxidant levels may play a part in the pathogenic cascade of many diseases, including AD (López et al. 2013; Swomley and Butterfield 2015). When the redox state of the cell is disturbed, it activates the oxidative stress-mediated signalling cascade, in which the cell may eventually die (Gambhir et al. 2022; Swomley and Butterfield 2015). Studies have shown the brain regions responsible for cognitive impairment and the regions with the most neuronal degeneration found in AD have the greatest oxidative damage (Keller et al. 2005). Besides, oxidative damage of essential biomolecules such as proteins, nucleic acids and lipids has been observed in multiple regions within the autopsied brain of patients with aMCI (López et al. 2013; Markesbery and Lovell 2007). The build-up of products from free radical damage in the brain of MCI subjects is also accompanied by a reduced enzymatic antioxidant defence, in which all these observations suggest that oxidative damage is an early event in the onset of AD (Padurariu et al. 2010).

In fact, aside from the aforementioned amyloid cascade hypothesis, the oxidative stress hypothesis has also been proposed as a pathogenic hypothesis for AD. This hypothesis seems even more convincing and attractive because it encompasses other etiologic mechanisms, such as the mitochondrial defect hypothesis, trace element toxicity hypothesis and amyloid cascade hypothesis (Lovell and Markesbery 2007; Markesbery and Lovell 2007). The presence of increased Aβ peptides was found to generate ROS in AD, mediating the oxidative damage to the biomolecules (Lovell and Markesbery 2007). It was suggested that the soluble Aβ oligomers are more toxic than the insoluble ones, and they exert the most oxidative damage to the cell, possibly owing to their ability of membrane translocation (Swomley and Butterfield

2015). Meanwhile, the mitochondria are also a factory of oxidant production. As a part of the normal functioning of mitochondria, there is a leak of O_2^- throughout the oxidative phosphorylation process, which is caused by the imperfect efficiency of electron transfer at the electron transport chain (Swomley and Butterfield 2015). The role of mitochondrial dysfunction in MCI will be further discussed in the following content.

Lipid peroxidation is considered a highly important mechanism throughout oxidative damage in the brain because of the brain's high lipid composition (López et al. 2013). The process can be divided into initiation, propagation and termination (Ayala et al. 2014). The initiation and propagation stages involve allylic hydrogen abstraction and radical chain reactions, which can spread between membrane systems (Swomley and Butterfield 2015). The propagation of chain reactions will only eventually be terminated by the interference of an antioxidant, in which it donates a hydrogen atom to the lipid peroxyl radical and undergoes reactions to form non-radical products at the end (Ayala et al. 2014).

Lipid peroxidation can result in structural membrane damage, and it yields the decomposed hydroperoxide by-products, in which some of them are the secondary bioreactive aldehydes that are toxic to the neurons, such as acrolein and 4-hydroxy-2-nonenal (4-HNE) (Bradley-Whitman and Lovell 2015; Markesbery and Lovell 2007; Wang et al. 2006). Another category of by-products from cerebral peroxidation is the products of oxygenated lipid rearrangement, which includes neuroprostanes (NeuroPs) and isoprostanes (IsoPs) (Bradley-Whitman and Lovell 2015). NeuroPs and IsoPs are non-toxic but useful as markers for DHA and arachidonic acid (ARA) oxidation, as they are derived from DHA and ARA, respectively (Bradley-Whitman and Lovell 2015; Lovell and Markesbery 2007; Markesbery and Lovell 2007). These by-products are produced in situ, stored in the membrane and may be secreted upon hydrolysis (Bradley-Whitman and Lovell 2015). One of the earliest studies has reported higher levels of 8,12-iso-iPF2α-VI in CSF, urine and plasma in elderly with MCI compared to elderly with normal cognitive function (Praticò et al. 2002). During the follow-up in this study, five out of 33 MCI subjects with high 8,12-iso-iPF2α-VI levels progressed to AD.

Protein dysfunction is another concern resulting from oxidative stress. As the proper function of protein relies on its structural integrity, oxidatively modified proteins can lose their function due to the conformational changes or unfolding process in oxidation (Swomley and Butterfield 2015). This causes the hydrophobic amino acids to be exposed to the aqueous environment, which leads to an aggregation of oxidised protein and forms cytoplasmic inclusions (Sultana et al. 2009; Swomley and Butterfield 2015). Oxidatively modified proteins may disrupt cellular functions in various ways, such as by altering gene regulation, inducing apoptosis and necrosis and changing protein turnover (Sultana et al. 2009). Protein carbonylation is the most common type of direct protein oxidation, and a higher concentration of protein carbonyls is detected in superior and middle temporal gyri in MCI patients compared to normal individuals (Markesbery and Lovell 2007; Swomley and Butterfield 2015). Protein oxidation can also occur indirectly, in which the protein is modified upon interacting with secondary by-products of oxidative stress, for example, acrolein or 4-HNE produced from lipid peroxidation (Swomley and Butterfield 2015). Through the process of Michael addition, these α/β-unsaturated aldehydes can become attached to proteins and ultimately change the protein activity (Swomley and Butterfield 2015).

2.3.2 DNA Damage and MCI

Ranging from products of cell metabolism to environmental chemicals, DNA in the human body cells is constantly attacked by agents from various endogenous and exogenous sources (Madabhushi et al. 2014). Damage to DNA is considered particularly harmful among all biological macromolecules as it serves as the blueprint for protein synthesis, and thus, DNA repair pathways such as the base excision repair (BER), nucleotide excision repair (NER) and nonhomologous end joining (NHEJ) are essential to prevent the potentially deleterious effects caused by the lesions (Madabhushi et al. 2014; Narciso et al. 2016).

While certain types of body cells have a relatively short lifespan and are continuously replaced, neurons need to endure this damage for a lifetime (Madabhushi et al. 2014). Linking to the previous discussion on oxidative stress in MCI, free radical-mediated DNA alterations have been found to pose a significant threat to the genome stability of mature neurons (Narciso et al. 2016). These alterations include DNA–DNA and DNA–protein crosslinking, DNA base modification and DNA strand breaks (Lovell and Markesbery 2007; Markesbery and Lovell 2007). As the transcription and protein translation are affected, it can result in neuronal death as the diminished proteins compromise neuron function (Markesbery and Lovell 2007).

Numerous oxidised base adducts were found to result from ROS attacks. Due to the low oxidation potential of guanine, it is most readily oxidised or most sensitive to the attack (Flora 2014; Lovell and Markesbery 2007). One of the most abundant modifications detected in urine excretion is 8-hydroxyguanine (8-OHG), which is an adduct formed via the C8 hydroxylation of guanosine (Flora 2014; Lovell and Markesbery 2007). Because of its easy collection, it is now serving as a predominant biomarker for oxidative DNA damage (Lovell and Markesbery 2007). Generally, mitochondrial DNA (mtDNA) has a higher susceptibility to free radical-mediated damage, owing to its poor repair capacity, a lack of protective histones and the proximity to ROS production in the respiratory chain (Lovell and Markesbery 2007). Wang et al. (2006) have found a significant increase of 8-OHG in nuclear DNA from the frontal and temporal lobe and in mtDNA in the temporal lobe among the MCI patients, as compared to the age-matched control groups, reaffirming the contribution of oxidative DNA damage in the neurodegeneration pathology.

In addition, single-strand breaks (SSBs) are the major manifestation of genomic injury, which are produced via a ROS attack on the DNA backbone, causing the phosphodiester bonds to break in one strand (Welch and Tsai 2022). Alternatively, SSBs can be generated indirectly, in which they appear as by-products of aborted topoisomerase I (TOP1) activity or intermediates of the BER pathway (Madabhushi et al. 2014). If the chromosomal SSBs are not repaired shortly, they will possibly block the DNA replication forks or cause them to collapse during the S phase of the cell cycle, forming double-strand breaks (DSBs) (Caldecott 2008). When there is a surge in SSBs, it may saturate the homologous recombination (HR) repair pathway and give rise to genetic instability or cell death. In post-mitotic neurons, which are non-dividing, SSBs may also impede the action of RNA polymerases in transcription (Caldecott 2008).

2.3.3 *Mitochondrial Dysfunction and MCI*

Mitochondria are a major power source that supplies ATP via oxidative phosphorylation, which is especially needed for energy-taxing neuronal activities (Wang et al. 2020a). Good mitochondrial health is essential to provide energy for endogenous neuroprotective and reparative mechanisms (Sharma et al. 2021). To keep the mitochondria functioning normally, a proper electrochemical gradient has to be maintained, which is greatly dependent upon the intactness of the mitochondrial structure (Wang et al. 2014). As unveiled by electron microscopy, loss of the internal structure and broken cristae were the characteristics of the vulnerable neurons in biopsied brain tissues from AD patients (Wang et al. 2014). Mitochondrial dysfunction has been shown to be involved in various hallmark features of AD, including the formation of neurofibrillary tangles, Aβ aggregates, neuroinflammation, oxidative stress and impaired synaptic transmission and plasticity (Sharma et al. 2021).

The function of mitochondria can be debilitated by several factors. While ageing changes their morphology and function, mtDNA is also prone to mutations and exposure to environmental toxins such as industrial toxic waste, pesticides and heavy metals have been shown to cause neurotoxicity through mitochondrial dysfunction (Sharma et al. 2021). When mitochondria are damaged, not only the energy metabolism and ATP production are affected, but they also lose the capability to undergo fission or fusion efficiently and are impaired for trafficking (Dagda 2018). These disturbances can bring about neuronal degeneration and malfunction, which are linked to calcium dyshomeostasis and excessive free radical production (Sharma et al. 2021).

Mitochondrial fission and fusion occur in the cytoplasm from time to time to regulate the distribution and morphology of the organelle (Wang et al. 2014). While fusion mediates the process of compensating the dysfunctional components of damaged mitochondria using functional constituents from healthy organelles, fission refers to fragmentation, in which the dysfunctional organelles are segregated from the mitochondrial network and reduced to a size that would facilitate autophagosomal encapsulation (Norat et al. 2020). A disequilibrium between fusion and fission can result in several events such as morphological alterations, depolarisation and mitochondrial swelling, which may add to susceptibility to other types of neuronal stress (Sharma et al. 2021). For example, an enhanced fusion or inhibited fission can cause the formation of large mitochondria whereby excessive fission may cause structural damage in vulnerable neurons in the AD brain (Wang et al. 2014).

In the AD neurons, a change in size and number of mitochondria suggests a fragmented mitochondrial network in the brain and is considered some of the abnormal mitochondrial dynamics in AD pathogenesis (Sharma et al. 2021; Wang et al. 2020a). Besides, defective fission or fusion activities also indirectly increase ROS formation via some negative effects on mtDNA integrity, calcium handling and bioenergetics (Wang et al. 2014). Specifically, there will be a drop in the neurons' calcium buffering capacity and rapid build-up of mtDNA mutations when the enhanced mitochondrial fission is unregulated (Wang et al. 2014).

As the predominance of fission over fusion can increase oxidative stress in AD and accelerate deterioration, inhibiting mitochondrial fission or promoting mitochondrial fusion would efficiently prevent ROS overproduction (Sharma et al. 2021; Wang et al. 2014). Mitochondrial fission also elicits apoptosis; therefore, blocking fission will help impede apoptosis in neurons (Dagda 2018). On the other hand, the application of ROS may disrupt the balance between mitochondrial fission and fusion, leading to further mitochondrial dysfunction and subsequently resulting in ROS overproduction, forming a vicious cycle (Wang et al. 2014).

Impaired calcium homeostasis is another important consequence of dysfunctional mitochondria. This is because mitochondria provide buffering machinery to modulate calcium concentration for signal transduction and also supply energy for the transmembrane Ca^{2+} pumps in the cell membrane and endoplasmic reticulum (Sharma et al. 2021; Wang et al. 2020a). Inadequate mitochondrial buffering can elevate the intracellular Ca^{2+} and impair neurotransmitter release and synaptic communication, causing neuronal loss (Sharma et al. 2021). Calcium dysregulation has also been suggested to be an upstream event for Aβ aggregation and tau hyperphosphorylation (Sharma et al. 2021). While excessive Ca^{2+} in the cytosol may cause accumulation of Aβ, mitochondrial damage can also result from the actions of Aβ, in which they enter the synaptic mitochondria and impair synaptic function. As the energy demands at synapses cannot be fulfilled, neurotransmission will be impaired and eventually cognitive failure occurs (Reddy and Beal 2008).

2.3.4 Telomere Shortening and MCI

During cell division, genomic stability is maintained by the evolutionarily conserved DNA sequence at the end of chromosomes, namely telomere (Wang et al. 2020b). Each time a cell divides, this protective cap will be shortened as the repeated nucleotide sequences will not be entirely replicated (Scarabino et al. 2017). Meanwhile, telomerase is a cellular ribonucleoprotein enzyme complex that acts to compensate for the loss of these sequences to maintain the telomere length, but its activity is usually low in somatic cells (Scarabino et al. 2017). As the telomeres are shortened progressively and depleted, cells will no longer reproduce and eventually die, causing tissue degeneration (Wang et al. 2020b).

Telomere shortening has been widely studied for its role in ageing, and such event has also been related to the pathogenesis of various diseases, such as neurological disorders (Cai et al. 2013; Wang et al. 2020b). A recent study has reported a significantly higher probability of AD progression for the lowest telomere length quartile group compared to the highest, among the MCI subjects with amyloid pathology, suggesting a link between rapid cognitive impairment and AD conversion from MCI (Koh et al. 2020). Risk factors for shortened telomeres include inflammation, immunologic response and replication stress and oxidative stress (Boccardi et al. 2015; Lee et al. 2020). Leucocyte telomere length (LTL) was suggested to reveal cellular age and has been widely used as a surrogate marker of telomere length in other organs (Eitan et al. 2014; Scarabino et al. 2017). Even though cell turnover for brain tissues is almost absent and thus telomere trimming may not even take place, shortening of leucocyte telomeres has been found associated with cognitive impairment (Scarabino et al. 2017).

On the other hand, between-person variation was observed in leucocyte telomere shortening during ageing, with sex and ethnicity as the most widely accepted factors that influence LTL (Aviv 2012; Boccardi et al. 2015; Eitan et al. 2014; Müezzinler et al. 2013). Moreover, the environment and genetics can interact to affect LTL shortening regardless of the status of neurodegenerative diseases (Eitan et al. 2014). The role of environmental factors was further supported by an interesting finding, in which similar telomere length has been observed for spouses, who generally may share similar diet, exercise habits or other environmental exposures, with a higher similarity seen in long-term couples (Eitan et al. 2014; Fletcher 2018). Therefore, LTL measurement may not necessarily provide a clear insight into the role of telomere shortening in cognitive decline, and it remains an unresolved question whether LTL is a useful indicator of the central nervous system (CNS) telomere length (Eitan et al. 2014).

Aside from the argument on the use of LTL as a biomarker, contradicting results have also been found on the relationship between telomere length and neurodegeneration, which may be explained by the diverse causes of telomere shortening (Eitan et al. 2014). While many findings showed a rapid telomere erosion in neurodegenerative diseases, there are studies claiming that telomere length does not predict cognitive impairment (Lee et al. 2020; Wang et al. 2020b). In a longitudinal study, there was no evidence found for the use of telomere length to distinguish the oldest old with normal cognition, dementia and MCI (Zekry et al. 2010). Not only that telomere length not associated with any change in cognitive status after a two-year follow-up, but the study has also not found any significant difference in telomere length among subjects with different aetiologies or severities of dementia (Zekry et al. 2010). Besides, a review has concluded a non-consistent relationship between LTL with AD and Parkinson's disease (PD), in which half of the studies included in the review reported no change in LTL and another half reported LTL shortening (Eitan et al. 2014). Surprisingly, evidence also exists for the role of long telomere lengths in the pathogenesis of MCI. In an elderly cohort that involved participants from the Mayo Clinic Study of Aging, subjects with the shortest and longest telomeres were shown to have an elevated risk of aMCI (Roberts et al. 2014). In all, the contribution of telomere shortening as part of the pathogenesis of AD or MCI is still unclear because of the variability between studies and further study with telomere length measurements in not only leucocytes but also other tissues may be useful to provide better insights on this topic.

2.4 Trace Elements and Their Role in Cognitive Function

2.4.1 Calcium

As a key signalling molecule in neurons, calcium acts to regulate neuronal gene expression, synaptic transmission, membrane excitability and other processes that are linked to memory and learning (Kawamoto et al. 2012; Meramat et al. 2015). A well-controlled calcium homeostasis is essential to maintain neuronal integrity and support brain physiology (Marambaud et al. 2009). In cells, calcium signals can arise upon electrical or receptor-mediated stimulation to trigger extracellular calcium influx, via the opening of specific calcium channels such as the voltage-gated calcium channels (VGCCs) and N-methyl-D-aspartate receptors (NMDARs) (Marambaud et al. 2009). Calcium can also be released from the intracellular calcium pool into the cytosol, predominantly from the endoplasmic reticulum (ER), through activation of ryanodine receptors (RyRs) or the inositol trisphosphate receptors (IP3Rs) (Hidalgo and Núñez 2007). Meanwhile, the mechanism of store-operated calcium entry (SOCE) is important in the replacement of ER calcium levels in non-excitable cells (Ge et al. 2022). Along with SOCE, sarcoplasmic-ER ATPase (SERCA) maintains the basal cytosolic calcium levels by regulating an active uptake of calcium into the internal stores (Marambaud et al. 2009).

Followed by the transient rise in calcium concentration, the calcium signal is then propagated to the nucleus. More exactly, a cascade of signalling events will be activated to bring about gene expression, which involves the activation of cAMP response element binding protein (CREB) and phosphorylation at Ser133 (Marambaud et al. 2009). As the nuclear calcium levels increase, gene programmes are activated to cause functional and structural changes within the cell and also the network it belongs to. These alterations can include a continual increase of synaptic efficacy or the build-up of a neuroprotective shield, which is thought to be a cellular correlate of learning and memory (Bading 2013).

While an increased intracellular calcium level is required for normal brain function, a small but sustained rise in calcium can ultimately cause neuronal cell death and memory disturbance (Meramat et al. 2015; Schram et al. 2007). There is growing evidence that disrupted calcium homeostasis contributes to the pathogenesis of cognitive decline and that calcium is closely linked to hallmarks of AD pathology like abnormal synaptic plasticity, Aβ deposition and hyperphosphorylation of tau protein (Ge et al. 2022; Rossom et al. 2012; Schram et al. 2007). Seemingly, most ER calcium-related proteins are associated with Aβ formation. For instance, inhibition of the SERCA pump inhibits Aβ production, whereby its overexpression increases Aβ accumulation (Jaworska et al. 2013). Besides, the knockdown of the IP3R also reduces Aβ generation, while the expression of RyR does the opposite (Jaworska et al. 2013). Synaptic plasticity encompasses two major forms—long-term potentiation (LTP) and long-term depression (LTD), which are both believed to be the mechanisms underlying learning, memory and amnesia (Ge et al. 2022). In the neurons of AD patients, the balance between LTP and LTD can be disrupted by high ER calcium (Ge et al. 2022). Additionally, increased intracellular calcium also leads to excessive calcium release and obstructs the induction of LTP (Ge et al. 2022).

Since Ca^{2+} diffuses through the BBB easily, serum calcium level is directly linked to the extracellular calcium in the brain (Schram et al. 2007). Results from the Rotterdam Study and the Leiden 85-plus Study have supported a negative effect of high serum calcium on cognitive function, in which an association was demonstrated between high serum calcium and a faster cognitive decline for individuals aged 75 and above (Schram et al. 2007). Conversely, low serum calcium has been found to correlate with a higher risk of MCI-to-AD conversion in a Japanese study cohort, but the underlying cause is still unclear (Sato et al. 2019). Considering the multiple functional roles of calcium in the human body along with the health concern of high intraneuronal calcium, studies have also looked into the effects of calcium supplementation. Rossom et al. (2012) have not found any significant difference in incident MCI between elderly women who received calcium carbonate plus vitamin D3 supplementation and the placebo group, but future investigation on calcium and vitamin D separately would be needed as it is difficult to determine their effect when both nutrients act to influence neuronal functioning differently. In addition, a cross-sectional study did not support the importance of calcium in cognitive performance among adults, and it suggested that the interindividual variation in serum calcium concentration was not linked to cognitive function during early adulthood (Tolppanen et al. 2011).

2.4.2 Copper

Copper is a cofactor of various enzymes, which plays a crucial role in the biological redox process. It takes part in reactions such as energy metabolism, iron metabolism, antioxidative defence and the synthesis of neuropeptides and neurotransmitters (Grochowski et al. 2019; Scheiber et al. 2014). Most copper enters the brain parenchyma as a free ion. Upon crossing the BBB, copper will be released into the CSF and later absorbed by the choroid epithelial cells (Grochowski et al. 2019). Free copper (bound to light compounds loosely) is said to affect brain function rather than those tightly bound to ceruloplasmin, because of the permeability of the BBB that will only allow the passage of low molecular weight compounds (Meramat et al. 2015; Salustri et al. 2010).

The copper deposition has been found to be higher in substantia nigra and locus coeruleus (LC), which contain catecholaminergic cells and are brain structures that are known for their association with attention (Grochowski et al. 2019; Salustri et al. 2010). The neurotransmitter norepinephrine (NE) is mainly produced in LC, and copper serves as an essential cofactor for dopamine β-hydroxylase, an enzyme that participates in the final biosynthetic step of NE synthesis (Rihel 2018). In LC, neurons that express NE are a predominant positive regulator for arousal and wakefulness (Rihel 2018). The LC-NE system also partakes in other brain functions and processes such as memory, emotional regulation, sensory processing and autonomic activity (Van Egroo et al. 2022). Although the effect of copper deficiency on the NE production in LC in relation to the sleep–wake behaviour is yet to be understood, increasing evidence supports that fragmented sleep–wake cycle and shorter and shallower sleep during old age contribute to the pathology of neurodegenerative diseases, with the LC-NE system as a nexus in between (Van Egroo et al. 2022).

Mitochondrial efficiency is crucial for the brain since it generates the most ATP required in the organ. The role of copper in energy metabolism is highlighted by the final step of electron transfer in the respiratory chain in mitochondria. Copper is the prosthetic group of complex IV (CIV) cytochrome c oxidase, a protein that is needed to catalyse the oxidation of reduced cytochrome c (Scheiber et al. 2014). In addition, copper is a cofactor of Cu/Zn-SOD, which functions to relieve ROS generated from the respiratory chain (Ruiz et al. 2021). It converts O_2^- to dioxygen and H_2O_2 for further disposal via the action of enzymes including glutathione peroxidase, catalase and peroxiredoxins (Scheiber et al. 2014). Thus, copper deficiency can impair the activities of the copper-dependent cytochrome c oxidase as well as Cu/Zn-SOD, which may compromise the antioxidative defence and expose the brain to more sources of oxidative stress (Scheiber et al. 2014). Similar to copper, iron is responsible for multiple enzymatic processes in the brain and its metabolism has to be well-regulated. This makes copper also important for iron homeostasis as it is required for the proper function of ferroxidase ceruloplasmin, which is to oxidise ferrous iron into the ferric form (White et al. 2012).

Despite being an essential micronutrient, copper can become toxic when its level becomes overly high. Modified copper homeostasis has been observed in some neurodegenerative diseases, in which the alteration of copper metabolism in the brain appears to contribute to elevated oxidative stress, causing neuronal toxicity (Pham et al. 2013). As reported, AD neuropil has a Cu (II) level, which is 400% greater than that of a healthy brain (Sarell et al. 2009). Copper can accumulate in the human body not only through food but also via drinking water piped through copper plumbing (Salustri et al. 2010). The toxic effect of copper is owing to its potential as a catalyst for tissue oxidative damage via redox cycling between Cu(I) and Cu (II) (Pham et al. 2013). Such toxicity involves the Fenton-type redox reactions, which describe the formation of damaging OH• resulting from the reaction between Cu(I) and H_2O_2, a by-product of oxygen metabolism within cells (Pham et al. 2013). OH• are highly reactive and are the most potent of free radicals generated; thus, copper toxicity will ultimately cause oxidative cell damage and death (Meramat et al. 2015; Pham et al. 2013). Nonetheless, there are also studies suggesting that a higher oxidation state of copper may be the active intermediate instead of OH•, under certain conditions (Pham et al. 2013).

Besides, another neurotoxic outcome resulting from excessive intracellular copper is the production of ROS via its interaction with APP and Aβ peptides (Mathys and White 2017). APP and Aβ peptide fragments have sequence motifs, which provide specific binding sites for copper (Purushothaman et al. 2020). Cu (II) binds Aβ peptides with high affinity, raising the proportions of α-helices and β-sheets in the peptides, which could worsen Aβ aggregation (Bagheri et al. 2018). Additionally, fibril formation greatly relies on pH and Cu (II) further encourages its formation by enabling it to take place at physiological pH (Bagheri et al. 2018). Studies have revealed several outcomes of the binding of Cu (II) to β-amyloid. In AD, the formation of dityrosine-linked dimers of β-amyloids is observed. Besides changing its conformation from parallel to anti-parallel, Cu (II) also acts to stabilise the structure. These dimers are prevented from degrading into monomers, and most importantly, they exert a neurotoxic property (Bagheri et al. 2018). Aside from Aβ, dyshomeostasis of copper may also further induce the formation of neurofibrillary tangles by interacting with the tau peptides (Meramat et al. 2015; Purushothaman et al. 2020).

2.4.3 Zinc

In the brain, zinc mainly serves as a component of metalloprotein, where they are bound to various enzymes, neuropeptides, hormones and hormone receptors to carry out structural or catalytic functions (Grochowski et al. 2019; Huskisson et al. 2007; Takeda 2001). Along with copper, it affects the antioxidant mechanism by forming an integral part of Cu/Zn-SOD to aid the efficient removal of O_2^- to sustain membrane integrity, as mentioned in the previous section (Meramat et al. 2015). Additionally, zinc appears to be a potent antioxidant to provide cellular defence against oxidative damage by interacting with thiols to protect them from oxidation and competing with metal ions that generate ROS (Delima et al. 2007; Olechnowicz et al. 2018).

Ionic zinc is also highly concentrated in synaptic vesicles of zinc-containing neurons, which exerts crucial effects on synaptic function and acts as a neuromodulator of various receptors such as γ-aminobutyric acid (GABA), α-amino-3-hydroxy-5-methyl-4-isoxazolepropionate (AMPA) and N-methyl-D-aspartate (NMDA) receptors (Grochowski et al. 2019; Huskisson et al. 2007; Takeda 2001). Zinc-containing neurons are considered a subclass of glutaminergic neurons, and their neuron terminals mainly exist in the hippocampal and amygdalar regions (Frederickson and Moncrieff 1994; Takeda 2001). Upon excitation, zinc and neurotransmitter glutamate will be co-released into the synaptic cleft of these neurons and modulate AMPA and NMDA receptors (Watt et al. 2010). Glutamate is an excitatory transmitter in the CNS that gives out rapid effects after it is sensed by the ionotropic and metabotropic receptors (Mocchegiani et al. 2005). Transmission of glutamate is vital for the functioning of the nervous system, but its long-lasting increased levels within the synapse may cause cell death and its hyperactivity has been linked to neurodegenerative diseases (Pochwat et al. 2015). Therefore, the regulation of zinc homeostasis is particularly important as normal glutamate activity requires good control from this endogenous regulatory factor (Pochwat et al. 2015). With the presence of such a relationship between zinc and glutamate, it also means that zinc is responsible to regulate and influence the overall brain excitability and synaptic plasticity (Watt et al. 2010).

Other than glutamatergic neurotransmission, high levels of zinc release may also modulate GABA functions in the forebrain (Mocchegiani et al. 2005). On the other hand, most zinc-enriched terminals are GABAergic in the spinal cord and that zinc may have a direct postsynaptic modulation on GABA receptors (Mocchegiani et al. 2005). Furthermore, zinc can inhibit calcium release from neuronal internal stores by interacting with the group I metabotropic glutamate receptors and influencing the voltage-gated calcium channels, which again relates to the disruption of calcium homeostasis that has been linked to cognitive impairment (Mocchegiani et al. 2005; Petrilli et al. 2017).

When zinc is deficient, zinc homeostasis is dysregulated in regions with vesicular zinc such as the hippocampus, which could lead to olfactory dysfunction and learning impairment (Takeda 2001). Zinc deficiency also increases the susceptibility to epileptic seizures, further highlighting the role of zinc in neuronal activity (Takeda 2001). Considering its function in antioxidation, inadequate zinc consumption is also postulated to compromise oxidative stress defence and DNA repair, exposing the cells to higher chances of DNA damage (Meramat et al. 2015). With the vulnerability of the brain to oxidative stress, cognitive function can be impaired and may result in abnormalities in brain function (Meramat et al. 2015).

Similar to copper, zinc not only binds Aβ to promote plaque formation but also accelerates Tau peptide aggregation (Purushothaman et al. 2020). Under in vitro conditions, Aβ peptides can aggregate into protease-resistant deposits when zinc ions are present (Purushothaman et al. 2020; Watt et al. 2010). Its role in AD pathogenesis has been supported by the observation of increased zinc in AD patients' neuropil as well as zinc enrichment in AD plaques (Purushothaman et al. 2020). However, zinc may also provide a protective effect against Aβ toxicity, possibly competing with copper for the Aβ binding sites to prevent free radical production (Purushothaman et al. 2020). It appears that the effect of zinc is in fact dependent on the concentration, in which a low zinc level promotes neuroprotection whereby a high concentration promotes Aβ toxicity (Rezaei-Ghaleh et al. 2011).

2.4.4 Selenium

The importance of selenium in cognitive function is apparent via the high priority of the brain to retain this element. When dietary selenium is deficient, selenium levels in the brain are maintained at the expense of other organs and irreversible brain injury can result as it depletes (Chen and Berry 2003; Rayman 2012). Selenium is extensively studied for its antioxidant role, particularly in the activity of glutathione peroxidase (GPxs) (Chen and Berry 2003). GPxs are capable of reducing the harmful peroxides to H_2O or alcohol to protect the unsaturated lipids or cell membranes in the brain (Chen and Berry 2003). As an important enzyme for antioxidant defence, the synthesis of GPxs is prioritised when the supply of selenium runs low (Rayman 2012). Besides, the expression of GPxs has been seen to increase surrounding the damaged region in occlusive cerebrovascular disease and PD, which supports its protective role in oxidation (Chen and Berry 2003).

GPx is in fact one of the selenoproteins, which are referred to as the proteins formed via the incorporation of selenocysteine at their active sites (Naderi et al. 2021). Most biological functions of selenium are believed to be exerted by selenoproteins (Chen and Berry 2003; Naderi et al. 2021). Aside from GPxs, enzymes such as methionine sulphoxide reductases, thioredoxin reductases and phospholipid hydroperoxide glutathione peroxidase are also involved in immunomodulation and free radical scavenging (Naderi et al. 2021). Furthermore, the neuroprotective mechanism of selenium has been found to involve Ca^{2+} influx modulation and anti-inflammatory pathways (Solovyev 2015). For instance, an ER-resident selenoprotein, SELENOK, was shown to influence Ca^{2+} influx by regulating IP3R palmitoylation and its overexpression can cause an elevation of IP3R-mediated free Ca^{2+} level in microglia (Zhang and Song 2021). Besides, overexpression of another selenoprotein, SELENOM, has been observed to decrease intracellular Ca^{2+} flux that is mediated by H_2O_2 in the neuronal cells (Zhang and Song 2021). Meanwhile, when the SELENOM gene is knocked out, it raises the cytosolic Ca^{2+} levels and enhances the event of oxidative stress and apoptosis (Zhang and Song 2021).

SELENOP is a well-known selenoprotein that is highly expressed and has multiple roles in the brain including selenium transport, storage and metabolism (Naderi et al. 2021; Zhang and Song 2021). For example, SELENOP delivers selenium to the neurons by binding to apoER2, a member of the lipoprotein (Rayman 2012). Other than its antioxidant activity, SELENOP has also been reported to promote the survival of neuronal cells (Chen and Berry 2003). It is able to inhibit Aβ aggregation and reduce the neurotoxicity of these Aβ metal complexes by competing Aβ for metal ions or forming complexes with these Aβ and metal ions. (Solovyev 2020; Zhang and Song 2021). However, the interaction between selenium and Aβ may lead to selenium restriction, which deprives selenium in the brain and reduces selenoprotein production (Solovyev 2020). SELENOP also seems to interact with tau proteins and play a part in cytoskeleton assembly, even though such interactions were mostly investigated in vitro (Solovyev 2020; Zhang and Song 2021).

Additionally, selenium is crucial for healthy ageing among the elderly population and that poor selenium status is highly associated with cognitive decline (Meramat et al. 2015). A cross-sectional survey has revealed a significant association between lower nail selenium levels and lower cognitive scores among rural Chinese elderly (Gao et al. 2007). This study involves a majority of subjects that have lived in the same area throughout their life, suggesting that a lifelong low exposure to selenium impairs brain function (Gao et al. 2007). Besides, Rita Cardoso et al. (2014) have demonstrated a decreased erythrocyte selenium concentration in elderly with AD and MCI when compared to the control group, which also supports the contribution of selenium deficiency on cognitive impairment among older people.

Nonetheless, it was pointed out that the results obtained from the studies that explored selenium involvement in neurodegeneration were difficult to compare (Solovyev 2015). Despite only involving the small case and control groups, many human studies also did not specify the selection criteria for the participants (Solovyev 2015). Similar to other trace elements discussed, overexposure to selenium may also bring deleterious effects on the CNS (Naderi et al. 2021). Excessive selenium can affect the regulation of cognitive functions by compromising the normal actions of proteins, signalling molecules and neurotransmitter systems involved, posing a risk for the development of neurodegenerative diseases (Naderi et al. 2021). For instance, selenium toxicity is suspected to contribute to the pathogenesis of amyotrophic lateral sclerosis, a progressive neuromuscular disease, by promoting mitochondrial abnormalities and oxidative damage (Naderi et al. 2021). Figure 2.1 summarises the relationship between dysregulation of trace elements and its mechanism associated with cognitive decline.

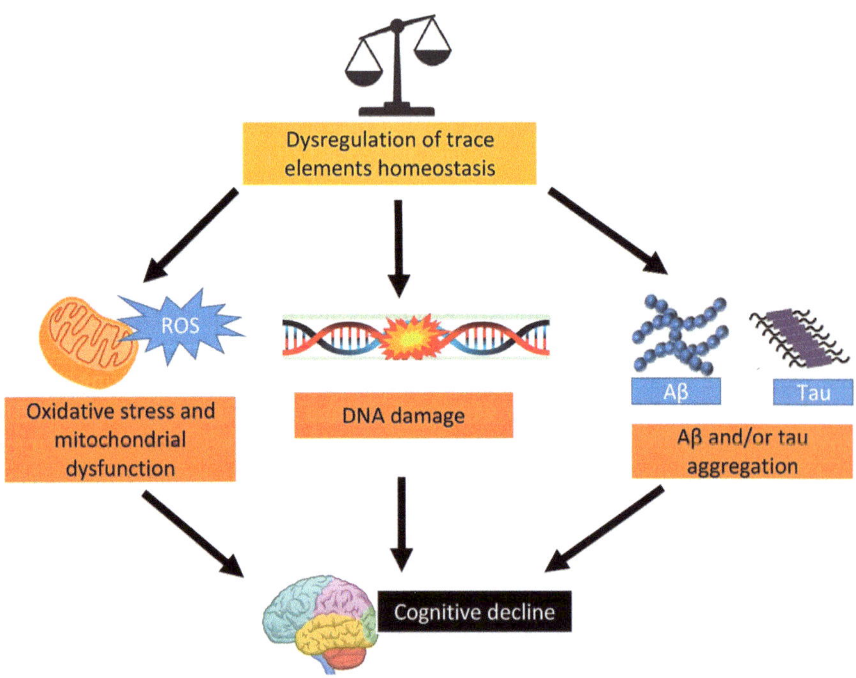

Fig. 2.1 Dysregulation of trace elements and its mechanism associated with cognitive decline. Trace element deficiencies and overexposure both result in oxidative stress, mitochondrial dysfunction, DNA damage, Aβ and/or tau aggregation, which further deteriorate neuron function and lead to cognitive decline

2.5 Conclusion

Trace elements form part of the proteins that participate in antioxidant protection of the CNS, and they regulate metabolic pathways that have been shown to change during ageing. A deficiency of these micronutrients is associated with impaired neuronal function, and it increases the risk of age-related disorders when a such deficiency occurs in a long term. Elements such as copper, zinc and selenium are particularly important due to their role in antioxidative defence within the brain, since oxidative stress and DNA damage are two of the most significant mechanisms involved in neurodegeneration. While the intake of trace elements is essential to modulate biological ageing, overexposure can have a detrimental effect. Disrupted homeostasis of the trace elements has also been linked with Aβ and/or tau aggregation, which can further contribute to oxidative stress. However, some protective effects were also found for zinc and selenium via their interaction with Aβ peptides. Meanwhile, studies on supplementation have provided mixed results, which may imply that cognitive decline is caused by the imbalance of multiple trace elements as they interact with one another. As the effect of long-term inadequacy or exposure to trace elements in cognition has not been thoroughly explored, further research is warranted to offer direct evidence to describe the link between these metals and MCI and to develop interventions to delay the onset of neurodegenerative diseases.

References

Alzheimer's Association. (2021) *Tau.* https://www.alz.org/media/Documents/alzheimers-dementia-tau-ts.pdf

Alzheimer's Association. (2022a). *Alzheimer's disease facts and figures.* https://www.alz.org/alzheimers-dementia/facts-figures

Alzheimer's Association. (2022b) *Mild cognitive impairment (MCI).* https://www.alz.org/alzheimers-dementia/what-is-dementia/related_conditions/mild-cognitive-impairment

Alzheimer's Society. (2022) *Mild cognitive impairment (MCI).* https://www.alzheimers.org.uk/about-dementia/types-dementia/mild-cognitive-impairment-mci

American Association for Clinical Chemistry. (2020) *Crucial evidence about APOE4's role in Alzheimer's disease.* https://www.aacc.org/cln/cln-stat/2020/may/21/crucial-evidence-about-apoe4s-role-in-alzheimers-disease

Aviv A (2012) Genetics of leukocyte telomere length and its role in atherosclerosis. *Mutat Res 730*(1-2):68–74. https://doi.org/10.1016/j.mrfmmm.2011.05.001

Ayala A, Muñoz MF, Argüelles S (2014) Lipid peroxidation: production, metabolism, and signaling mechanisms of malondialdehyde and 4-hydroxy-2-nonenal. *Oxid Med Cell Longev 2014*:360438. https://doi.org/10.1155/2014/360438

Bading H (2013) Nuclear calcium signalling in the regulation of brain function. *Nat Rev Neurosci 14*(9):593–608. https://doi.org/10.1038/nrn3531

Bagheri S, Squitti R, Haertlé T, Siotto M, Saboury AA (2018) Role of copper in the onset of Alzheimer's disease compared to other metals. *Front Aging Neurosci 9*:446. https://doi.org/10.3389/fnagi.2017.00446

Benson GS, Bauer C, Hausner L, Couturier S, Lewczuk P, Peters O, Hüll M, Jahn H, Jessen F, Pantel J, Teipel SJ, Wagner M, Schuchhardt J, Wiltfang J, Kornhuber J, Frölich L (2022) Don't forget about tau: the effects of ApoE4 genotype on Alzheimer's disease cerebrospinal fluid biomarkers in subjects with mild cognitive impairment-data from the Dementia Competence Network. *J Neural Transmission (Vienna, Austria : 1996) 129*(5-6):477–486. https://doi.org/10.1007/s00702-022-02461-0

Bhatti GK, Reddy AP, Reddy PH, Bhatti JS (2020) Lifestyle modifications and nutritional interventions in aging-associated cognitive decline and Alzheimer's disease. *Front Aging Neurosci 11*:369. https://doi.org/10.3389/fnagi.2019.00369

Blinkouskaya Y, Caçoilo A, Gollamudi T, Jalalian S, Weickenmeier J (2021) Brain aging mechanisms with mechanical manifestations. *Mech Ageing Dev 200*:111575. https://doi.org/10.1016/j.mad.2021.111575

Bloom GS (2014) Amyloid-β and tau: the trigger and bullet in Alzheimer disease pathogenesis. *JAMA Neurol 71*(4):505–508. https://doi.org/10.1001/jamaneurol.2013.5847

Boccardi V, Pelini L, Ercolani S, Ruggiero C, Mecocci P (2015) From cellular senescence to Alzheimer's disease: the role of telomere shortening. *Ageing Res Rev 22*:1–8. https://doi.org/10.1016/j.arr.2015.04.003

Bradley-Whitman MA, Lovell MA (2015) Biomarkers of lipid peroxidation in Alzheimer disease (AD): an update. *Arch Toxicol 89*(7):1035–1044. https://doi.org/10.1007/s00204-015-1517-6

Brion JP, Anderton BH, Authelet M, Dayanandan R, Leroy K, Lovestone S, Octave JN, Pradier L, Touchet N, Tremp G (2001) Neurofibrillary tangles and tau phosphorylation. *Biochem Soc Symp* 67:81–88. https://doi.org/10.1042/bss0670081

Broglio SP, Eckner JT, Paulson HL, Kutcher JS (2012) Cognitive decline and aging: the role of concussive and subconcussive impacts. *Exerc Sport Sci Rev 40*(3):138–144. https://doi.org/10.1097/JES.0b013e3182524273

Cabral Pinto M, Marinho-Reis P, Almeida A, Pinto E, Neves O, Inácio M, Gerardo B, Freitas S, Simões MR, Dinis PA, Diniz L, Ferreira da Silva E, Moreira PI (2019) Links between cognitive status and trace element levels in hair for an environmentally exposed population: a case study in the surroundings of the estarreja industrial area. *Int J Environ Res Public Health 16*(22):4560. https://doi.org/10.3390/ijerph16224560

Cai Z, Yan LJ, Ratka A (2013) Telomere shortening and Alzheimer's disease. *Neuromolecular Med 15*(1):25–48. https://doi.org/10.1007/s12017-012-8207-9

Caldecott KW (2008) Single-strand break repair and genetic disease. *Nat Rev Genet 9*(8):619–631. https://doi.org/10.1038/nrg2380

Catlow BJ, Jalloh A, Sanchez-Ramos J (2016) Hippocampal neurogenesis: effects of psychedelic drugs. *Neuropathology of Drug Addictions and Substance Misuse* 821-831. https://doi.org/10.1016/B978-0-12-800212-4.00077-7

Chen GF, Xu TH, Yan Y, Zhou YR, Jiang Y, Melcher K, Xu HE (2017) Amyloid beta: structure, biology and structure-based therapeutic development. *Acta Pharmacol Sin 38*(9):1205–1235. https://doi.org/10.1038/aps.2017.28

Chen J, Berry MJ (2003) Selenium and selenoproteins in the brain and brain diseases. *J Neurochem 86*(1):1–12. https://doi.org/10.1046/j.1471-4159.2003.01854.x

Chiu MJ, Chen YF, Chen TF, Yang SY, Yang FP, Tseng TW, Chieh JJ, Chen JC, Tzen KY, Hua MS, Horng HE (2014) Plasma tau as a window to the brain-negative associations with brain volume and memory function in mild cognitive impairment and early Alzheimer's disease. *Human Brain Mapp 35*(7):3132–3142. https://doi.org/10.1002/hbm.22390

Csukly G, Sirály E, Fodor Z, Horváth A, Salacz P, Hidasi Z, Csibri É, Rudas G, Szabó Á (2016) The differentiation of amnestic type MCI from the non-amnestic types by structural MRI. *Front Aging Neurosci 8*:52. https://doi.org/10.3389/fnagi.2016.00052

Dagda RK (2018) Role of mitochondrial dysfunction in degenerative brain diseases, an overview. *Brain Sci 8*(10):178. https://doi.org/10.3390/brainsci8100178

Dage JL, Wennberg A, Airey DC, Hagen CE, Knopman DS, Machulda MM, Roberts RO, Jack CR Jr, Petersen RC, Mielke MM (2016) Levels of tau protein in plasma are associated with neurodegeneration and cognitive function in a population-based elderly cohort. *Alzheimer's Dement 12*(12):1226–1234. https://doi.org/10.1016/j.jalz.2016.06.001

Dauncey MJ (2014) Nutrition, the brain and cognitive decline: insights from epigenetics. *Eur J Clin Nutr 68*(11):1179–1185. https://doi.org/10.1038/ejcn.2014.173

Del Parigi A, Panza F, Capurso C, Solfrizzi V (2006) Nutritional factors, cognitive decline, and dementia. *Brain Res Bull 69*(1):1–19. https://doi.org/10.1016/j.brainresbull.2005.09.020

Delima R, Trinder D, Olynyk JK (2007) Potential protective effects of zinc in iron overload. *Liver Int 27*(1):4–5. https://doi.org/10.1111/j.1478-3231.2006.01428.x

Demurtas J, Schoene D, Torbahn G, Marengoni A, Grande G, Zou L, Petrovic M, Maggi S, Cesari M, Lamb S, Soysal P, Kemmler W, Sieber C, Mueller C, Shenkin SD, Schwingshackl L, Smith L, Veronese N, European Society of Geriatric Medicine Special Interest Group in Systematic Reviews and Meta-Analyses, Frailty, Sarcopenia, and Dementia (2020) Physical activity and exercise in mild cognitive impairment and dementia: an umbrella review of intervention and observational studies. *J Am Med Dir Assoc 21*(10):1415–1422. e6. https://doi.org/10.1016/j.jamda.2020.08.031

Dominguez LJ, Barbagallo M (2018) Nutritional prevention of cognitive decline and dementia. *Acta Bio-Medica : Atenei Parmensis 89*(2):276–290. https://doi.org/10.23750/abm.v89i2.7401

Dugger BN, Davis K, Malek-Ahmadi M, Hentz JG, Sandhu S, Beach TG, Adler CH, Caselli RJ, Johnson TA, Serrano GE, Shill HA, Belden C, Driver-Dunckley E, Caviness JN, Sue LI, Jacobson S, Powell J, Sabbagh MN (2015) Neuropathological comparisons of amnestic and nonamnestic mild cognitive impairment. *BMC Neurol 15*:146. https://doi.org/10.1186/s12883-015-0403-4

Eitan E, Hutchison ER, Mattson MP (2014) Telomere shortening in neurological disorders: an abundance of unanswered questions. *Trends Neurosci 37*(5):256–263. https://doi.org/10.1016/j.tins.2014.02.010

Ellison JM, Harper DG, Berlow Y, Zeranski L (2008) Beyond the "C" in MCI: noncognitive symptoms in amnestic and non-amnestic mild cognitive impairment. *CNS Spectr 13*(1):66–72. https://doi.org/10.1017/s1092852900016175

Eshkoor SA, Hamid TA, Mun CY, Ng CK (2015) Mild cognitive impairment and its management in older people. *Clin Interv Aging 10*:687–693. https://doi.org/10.2147/CIA.S73922

Fletcher JM (2018) Spousal concordance in telomere length: new evidence from older adults in the US. *PloS One 13*(11):e0202388. https://doi.org/10.1371/journal.pone.0202388

Flora SJ (2014) Metals. *Biomarkers Toxicol* 485-519. https://doi.org/10.1016/b978-0-12-404630-6.00029-4

Frederickson CJ, Moncrieff DW (1994) Zinc-containing neurons. *Biol Signals 3*(3):127–139. https://doi.org/10.1159/000109536

Gambhir L, Tyagi G, Bhardwaj R, Kapoor N, & Sharma G (2022) Perturbation of cellular redox status: role of Nrf2, a master regulator of cellular redox. In (Ed.), Reactive oxygen species. IntechOpen. https://doi.org/10.5772/intechopen.102319

Gao S, Jin Y, Hall KS, Liang C, Unverzagt FW, Ji R, Murrell JR, Cao J, Shen J, Ma F, Matesan J, Ying B, Cheng Y, Bian J, Li P, Hendrie HC (2007) Selenium level and cognitive function in rural elderly Chinese. *Am J Epidemiol 165*(8):955–965. https://doi.org/10.1093/aje/kwk073

Ge M, Zhang J, Chen S, Huang Y, Chen W, He L, Zhang Y (2022) Role of calcium homeostasis in Alzheimer's disease. *Neuropsychiatr Dis Treat 18*:487–498. https://doi.org/10.2147/NDT.S350939

Goukasian N, Porat S, Blanken A, Avila D, Zlatev D, Hurtz S, Hwang KS, Pierce J, Joshi SH, Woo E, Apostolova LG (2019) Cognitive correlates of hippocampal atrophy and ventricular enlargement in adults with or without mild cognitive impairment. *Dementia Geriatr Cogn Disord Extra 9*(2):281–293. https://doi.org/10.1159/000490044

Grochowski C, Blicharska E, Krukow P, Jonak K, Maciejewski M, Szczepanek D, Jonak K, Flieger J, Maciejewski R (2019) Analysis of trace elements in human brain: its aim, methods, and concentration levels. *Front Chem 7*:115. https://doi.org/10.3389/fchem.2019.00115

Gu L, Guo Z (2013) Alzheimer's Aβ42 and Aβ40 peptides form interlaced amyloid fibrils. *J Neurochem 126*(3):305–311. https://doi.org/10.1111/jnc.12202

Hidalgo C, Núñez MT (2007) Calcium, iron and neuronal function. *IUBMB Life 59*(4-5):280–285. https://doi.org/10.1080/15216540701222906

Huskisson E, Maggini S, Ruf M (2007) The influence of micronutrients on cognitive function and performance. *J Int Med Res 35*(1):1–19. https://doi.org/10.1177/147323000703500101

Jäkel L, Boche D, Nicoll J, Verbeek MM (2019) Aβ43 in human Alzheimer's disease: effects of active Aβ42 immunization. *Acta Neuropathol Commun 7*(1):141. https://doi.org/10.1186/s40478-019-0791-6

Jaworska A, Dzbek J, Styczynska M, Kuznicki J (2013) Analysis of calcium homeostasis in fresh lymphocytes from patients with sporadic Alzheimer's disease or mild cognitive impairment. *Biochim Biophys Acta 1833*(7):1692–1699. https://doi.org/10.1016/j.bbamcr.2013.01.012

Jongsiriyanyong S, Limpawattana P (2018) Mild cognitive impairment in clinical practice: a review article. *Am J Alzheimer's Dis Other Demen 33*(8):500–507. https://doi.org/10.1177/1533317518791401

Kawamoto EM, Vivar C, Camandola S (2012) Physiology and pathology of calcium signaling in the brain. *Front Pharmacol 3*:61. https://doi.org/10.3389/fphar.2012.00061

Keller JN, Schmitt FA, Scheff SW, Ding Q, Chen Q, Butterfield DA, Markesbery WR (2005) Evidence of increased oxidative damage in subjects with mild cognitive impairment. *Neurology 64*(7):1152–1156. https://doi.org/10.1212/01.WNL.0000156156.13641.BA

Kihlstrom JF (2018) Unconscious cognition. *Reference Module In Neuroscience And Biobehavioral Psychology*. https://doi.org/10.1016/B978-0-12-809324-5.21860-9

Koh SH, Choi SH, Jeong JH, Jang JW, Park KW, Kim EJ, Kim HJ, Hong JY, Yoon SJ, Yoon B, Kang JH, Lee JM, Park HH, Ha J, Suh YJ, Kang S (2020) Telomere shortening reflecting physical aging is associated with cognitive decline and dementia conversion in mild cognitive impairment due to Alzheimer's disease. *Aging 12*(5):4407–4423. https://doi.org/10.18632/aging.102893

Law CK, Lam FM, Chung RC, Pang MY (2020) Physical exercise attenuates cognitive decline and reduces behavioural problems in people with mild cognitive impairment and dementia: a systematic review. *J Physiother 66*(1):9–18. https://doi.org/10.1016/j.jphys.2019.11.014

Le Gal K, Schmidt EE, Sayin VI (2021) Cellular redox homeostasis. *Antioxidants 10*(9):1377. https://doi.org/10.3390/antiox10091377

Lee EH, Han MH, Ha J, Park HH, Koh SH, Choi SH, Lee JH (2020) Relationship between telomere shortening and age in Korean individuals with mild cognitive impairment and Alzheimer's disease compared to that in healthy controls. *Aging 13*(2):2089–2100. https://doi.org/10.18632/aging.202206

Lo YL, Cheng SH (2023) Iron and Alzheimer's Disease. In: Mohamed W, Brogazzi NL, Kostrzewa RM (eds) Brain-iron cross talk. Nutritional neurosciences. Springer, Singapore. https://doi.org/10.1007/978-981-19-7327-7_7

Loh SP, Cheng SH, Mohamed W (2022) Edible bird's nest as a potential cognitive enhancer. *Front Neurol 13*:865671. https://doi.org/10.3389/fneur.2022.865671

López N, Tormo C, De Blas I, Llinares I, Alom J (2013) Oxidative stress in Alzheimer's disease and mild cognitive impairment with high sensitivity and specificity. *J Alzheimer's Dis 33*(3):823–829. https://doi.org/10.3233/JAD-2012-121528

Lopez OL (2013) Mild cognitive impairment. *Continuum (Minneapolis, Minn.) 19*(2 Dementia):411–424. https://doi.org/10.1212/01.CON.0000429175.29601.97

López-Sanz D, Garcés P, Álvarez B, Delgado-Losada ML, López-Higes R, Maestú F (2017) Network disruption in the preclinical stages of Alzheimer's disease: from subjective cognitive decline to mild cognitive impairment. *Int J Neural Syst 27*(8):1750041. https://doi.org/10.1142/S0129065717500411

Lovell MA, Markesbery WR (2007) Oxidative damage in mild cognitive impairment and early Alzheimer's disease. *J Neurosci Res 85*(14):3036–3040. https://doi.org/10.1002/jnr.21346

Madabhushi R, Pan L, Tsai LH (2014) DNA damage and its links to neurodegeneration. *Neuron 83*(2):266–282. https://doi.org/10.1016/j.neuron.2014.06.034

Marambaud P, Dreses-Werringloer U, Vingtdeux V (2009) Calcium signaling in neurodegeneration. *Mol Neurodegener 4*:20. https://doi.org/10.1186/1750-1326-4-20

Markesbery WR, Lovell MA (2007) Damage to lipids, proteins, DNA, and RNA in mild cognitive impairment. *Arch Neurol 64*(7):954–956. https://doi.org/10.1001/archneur.64.7.954

Martin E, Velayudhan L (2020) Neuropsychiatric symptoms in mild cognitive impairment: a literature review. *Dement Geriatr Cogn Disord 49*(2):146–155. https://doi.org/10.1159/000507078

Mathys ZK, White AR (2017) Copper and Alzheimer's disease. *Adv Neurobiol 18*:199–216. https://doi.org/10.1007/978-3-319-60189-2_10

Mattson MP, Arumugam TV (2018) Hallmarks of brain aging: adaptive and pathological modification by metabolic states. *Cell Metab 27*(6):1176–1199. https://doi.org/10.1016/j.cmet.2018.05.011

Mawuenyega KG, Kasten T, Sigurdson W, Bateman RJ (2013) Amyloid-beta isoform metabolism quantitation by stable isotope-labeled kinetics. *Anal Biochem 440*(1):56–62. https://doi.org/10.1016/j.ab.2013.04.031

McGrattan AM, McEvoy CT, McGuinness B, McKinley MC, Woodside JV (2018) Effect of dietary interventions in mild cognitive impairment: a systematic review. *Br J Nutr 120*(12):1388–1405. https://doi.org/10.1017/S0007114518002945

Meramat A, Rajab NF, Shahar S, Sharif R (2015) Cognitive impairment, genomic instability and trace elements. *J Nutr Health Aging 19*(1):48–57. https://doi.org/10.1007/s12603-014-0489-1

Metaxas A, Kempf SJ (2016) Neurofibrillary tangles in Alzheimer's disease: elucidation of the molecular mechanism by immunohistochemistry and tau protein phospho-proteomics. *Neural Regen Res 11*(10):1579–1581. https://doi.org/10.4103/1673-5374.193234

Mocchegiani E, Bertoni-Freddari C, Marcellini F, Malavolta M (2005) Brain, aging and neurodegeneration: role of zinc ion availability. *Progress Neurobiol 75*(6):367–390. https://doi.org/10.1016/j.pneurobio.2005.04.005

Monastero R, Mangialasche F, Camarda C, Ercolani S, Camarda R (2009) A systematic review of neuropsychiatric symptoms in mild cognitive impairment. *J Alzheimer's Dis 18*(1):11–30. https://doi.org/10.3233/JAD-2009-1120

Montagne A, Nation DA, Sagare AP, Barisano G, Sweeney MD, Chakhoyan A, Pachicano M, Joe E, Nelson AR, D'Orazio LM, Buennagel DP, Harrington MG, Benzinger T, Fagan AM, Ringman JM, Schneider LS, Morris JC, Reiman EM, Caselli RJ, Chui HC et al (2020) APOE4 leads to blood-brain barrier dysfunction predicting cognitive decline. *Nature 581*(7806):71–76. https://doi.org/10.1038/s41586-020-2247-3

Morris JC, Storandt M, Miller JP, McKeel DW, Price JL, Rubin EH, Berg L (2001) Mild cognitive impairment represents early-stage Alzheimer disease. *Arch Neurol 58*(3):397–405. https://doi.org/10.1001/archneur.58.3.397

Morris M (2012) Nutritional determinants of cognitive aging and dementia. *Proc Nutr Soc 71*(1):1–13. https://doi.org/10.1017/S0029665111003296

Müezzinler A, Zaineddin AK, Brenner H (2013) A systematic review of leukocyte telomere length and age in adults. *Ageing Res Rev 12*(2):509–519. https://doi.org/10.1016/j.arr.2013.01.003

Murman DL (2015) The impact of age on cognition. *Semin Hear 36*(3):111–121. https://doi.org/10.1055/s-0035-1555115

Naderi M, Puar P, Zonouzi-Marand M, Chivers DP, Niyogi S, Kwong R (2021) A comprehensive review on the neuropathophysiology of selenium. *Sci Total Environ 767*:144329. https://doi.org/10.1016/j.scitotenv.2020.144329

Narciso L, Parlanti E, Racaniello M, Simonelli V, Cardinale A, Merlo D, Dogliotti E (2016) The response to oxidative DNA damage in neurons: mechanisms and disease. *Neural Plast 2016*:3619274. https://doi.org/10.1155/2016/3619274

National Institute on Aging. (2017). *What happens to the brain in Alzheimer's disease?* https://www.nia.nih.gov/health/what-happens-brain-alzheimers-disease

Norat P, Soldozy S, Sokolowski JD, Gorick CM, Kumar JS, Chae Y, Yağmurlu K, Prada F, Walker M, Levitt MR, Price RJ, Tvrdik P, Kalani M (2020) Mitochondrial dysfunction in neurological disorders: exploring mitochondrial transplantation. *NPJ Regen Med 5*(1):22. https://doi.org/10.1038/s41536-020-00107-x

Ogawa S (2014) Nutritional management of older adults with cognitive decline and dementia. *Geriatr Gerontol Int 14(Suppl* 2):17–22. https://doi.org/10.1111/ggi.12252

Olechnowicz J, Tinkov A, Skalny A, Suliburska J (2018) Zinc status is associated with inflammation, oxidative stress, lipid, and glucose metabolism. *J Physiol Sci 68*(1):19–31. https://doi.org/10.1007/s12576-017-0571-7

Padurariu M, Ciobica A, Hritcu L, Stoica B, Bild W, Stefanescu C (2010) Changes of some oxidative stress markers in the serum of patients with mild cognitive impairment and Alzheimer's disease. *Neurosci Lett 469*(1):6–10. https://doi.org/10.1016/j.neulet.2009.11.033

Peters R (2006) Ageing and the brain. *Postgrad Med J 82*(964):84–88. https://doi.org/10.1136/pgmj.2005.036665

Petersen RC (2016) Mild cognitive impairment. *Continuum (Minneapolis, Minn.) 22*(2 Dementia):404–418. https://doi.org/10.1212/CON.0000000000000313

Petersen RC, Lopez O, Armstrong MJ, Getchius T, Ganguli M, Gloss D, Gronseth GS, Marson D, Pringsheim T, Day GS, Sager M, Stevens J, Rae-Grant A (2018) Practice guideline update summary: mild cognitive impairment: report of the guideline development, dissemination, and implementation Subcommittee of the American Academy of Neurology. *Neurology 90*(3):126–135. https://doi.org/10.1212/WNL.0000000000004826

Petrilli MA, Kranz TM, Kleinhaus K, Joe P, Getz M, Johnson P, Chao MV, Malaspina D (2017) The emerging role for zinc in depression and psychosis. *Front Pharmacol 8*:414. https://doi.org/10.3389/fphar.2017.00414

Pham AN, Xing G, Miller CJ, Waite TD (2013) Fenton-like copper redox chemistry revisited: hydrogen peroxide and superoxide mediation of copper-catalyzed oxidant production. *J Catal 301*:54–64. https://doi.org/10.1016/j.jcat.2013.01.025

Pochwat B, Nowak G, Szewczyk B (2015) Relationship between Zinc (Zn (2+)) and glutamate receptors in the processes underlying neurodegeneration. *Neural Plast 2015*:591563. https://doi.org/10.1155/2015/591563

Popa-Wagner A, Mitran S, Sivanesan S, Chang E, Buga AM (2013) ROS and brain diseases: the good, the bad, and the ugly. *Oxid Med Cell Longev 2013*:963520. https://doi.org/10.1155/2013/963520

Power R, Prado-Cabrero A, Mulcahy R, Howard A, Nolan J (2019) The role of nutrition for the aging population: implications for cognition and Alzheimer's disease. *Annu Rev Food Sci Technol 10*(1):619–639. https://doi.org/10.1146/annurev-food-030216-030125

Praticò D, Clark CM, Liun F, Rokach J, Lee VY, Trojanowski JQ (2002) Increase of brain oxidative stress in mild cognitive impairment: a possible predictor of Alzheimer disease. *Arch Neurol 59*(6):972–976. https://doi.org/10.1001/archneur.59.6.972

Purushothaman S, Ragavi A, Basavaraju P, Moorthi PV, Anand AV (2020) The role of trace elements in Alzheimer's disease. *Arch Mental Health 21*(2):107. https://doi.org/10.4103/amh.amh_18_20

Rayman MP (2012) Selenium and human health. *Lancet (London, England) 379*(9822):1256–1268. https://doi.org/10.1016/S0140-6736(11)61452-9

Reddy PH, Beal MF (2008) Amyloid beta, mitochondrial dysfunction and synaptic damage: implications for cognitive decline in aging and Alzheimer's disease. *Trends Mol Med 14*(2):45–53. https://doi.org/10.1016/j.molmed.2007.12.002

Rezaei-Ghaleh N, Giller K, Becker S, Zweckstetter M (2011) Effect of zinc binding on β-amyloid structure and dynamics: implications for Aβ aggregation. *Biophys J 101*(5):1202–1211. https://doi.org/10.1016/j.bpj.2011.06.062

Rihel J (2018) Copper on the brain. *Nat Chem Biol 14*(7):638–639. https://doi.org/10.1038/s41589-018-0089-1

Rita Cardoso B, Silva Bandeira V, Jacob-Filho W, Franciscato Cozzolino SM (2014) Selenium status in elderly: relation to cognitive decline. *J Trace Elem Med Biol 28*(4):422–426. https://doi.org/10.1016/j.jtemb.2014.08.009

Ritchie K (2004) Mild cognitive impairment: an epidemiological perspective. *Dialogues Clin Neurosci 6*(4):401–408. https://doi.org/10.31887/DCNS.2004.6.4/kritchie

Roberts RO, Boardman LA, Cha RH, Pankratz VS, Johnson RA, Druliner BR, Christianson TJ, Roberts LR, Petersen RC (2014) Short and long telomeres increase risk of amnestic mild cognitive impairment. *Mech Ageing Dev 141-142*:64–69. https://doi.org/10.1016/j.mad.2014.10.002

Rossom RC, Espeland MA, Manson JE, Dysken MW, Johnson KC, Lane DS, LeBlanc ES, Lederle FA, Masaki KH, Margolis KL (2012) Calcium and vitamin D supplementation and cognitive impairment in the women's health initiative. *J Am Geriatr Soc 60*(12):2197–2205. https://doi.org/10.1111/jgs.12032

Ruiz LM, Libedinsky A, Elorza AA (2021) Role of copper on mitochondrial function and metabolism. *Front Mol Biosci 8*:711227. https://doi.org/10.3389/fmolb.2021.711227

Salustri C, Barbati G, Ghidoni R, Quintiliani L, Ciappina S, Binetti G, Squitti R (2010) Is cognitive function linked to serum free copper levels? A cohort study in a normal population. *Clin Neurophysiol 121*(4):502–507. https://doi.org/10.1016/j.clinph.2009.11.090

Sanford AM (2017) Mild cognitive impairment. *Clin Geriatr Med 33*(3):325–337. https://doi.org/10.1016/j.cger.2017.02.005

Sarell CJ, Syme CD, Rigby SE, Viles JH (2009) Copper(II) binding to amyloid-beta fibrils of Alzheimer's disease reveals a picomolar affinity: stoichiometry and coordination geometry are independent of Abeta oligomeric form. *Biochemistry 48*(20):4388–4402. https://doi.org/10.1021/bi900254n

Sato K, Mano T, Ihara R, Suzuki K, Tomita N, Arai H, Ishii K, Senda M, Ito K, Ikeuchi T, Kuwano R, Matsuda H, Iwatsubo T, Toda T, Iwata A, Alzheimer's Disease Neuroimaging Initiative, and Japanese Alzheimer's Disease Neuroimaging Initiative (2019) Lower serum calcium as a potentially associated factor for conversion of mild cognitive impairment to early Alzheimer's disease in the Japanese Alzheimer's disease neuroimaging initiative. *J Alzheimer's Dis 68*(2):777–788. https://doi.org/10.3233/JAD-181115

Scarabino D, Broggio E, Gambina G, Corbo RM (2017) Leukocyte telomere length in mild cognitive impairment and Alzheimer's disease patients. *Exp Gerontol 98*:143–147. https://doi.org/10.1016/j.exger.2017.08.025

Scarmeas N, Stern Y, Mayeux R, Manly JJ, Schupf N, Luchsinger JA (2009) Mediterranean diet and mild cognitive impairment. *Arch Neurol 66*(2):216–225. https://doi.org/10.1001/archneurol.2008.536

Scheiber IF, Mercer JF, Dringen R (2014) Metabolism and functions of copper in brain. *Prog Neurobiol 116*:33–57. https://doi.org/10.1016/j.pneurobio.2014.01.002

Schram MT, Trompet S, Kamper AM, de Craen AJ, Hofman A, Euser SM, Breteler MM, Westendorp RG (2007) Serum calcium and cognitive function in old age. *J Am Geriatr Soc 55*(11):1786–1792. https://doi.org/10.1111/j.1532-5415.2007.01418.x

Sharma C, Kim S, Nam Y, Jung UJ, Kim SR (2021) Mitochondrial dysfunction as a driver of cognitive impairment in Alzheimer's Disease. *Int J Mol Sci 22*(9):4850. https://doi.org/10.3390/ijms22094850

Small GW, Kepe V, Ercoli LM, Siddarth P, Bookheimer SY, Miller KJ, Lavretsky H, Burggren AC, Cole GM, Vinters HV, Thompson PM, Huang SC, Satyamurthy N, Phelps ME, Barrio JR (2006) PET of brain amyloid and tau in mild cognitive impairment. *New Engl J Med 355*(25):2652–2663. https://doi.org/10.1056/NEJMoa054625

Solovyev N (2020) Selenoprotein P and its potential role in Alzheimer's disease. *Hormones (Athens, Greece) 19*(1):73–79. https://doi.org/10.1007/s42000-019-00112-w

Solovyev ND (2015) Importance of selenium and selenoprotein for brain function: From antioxidant protection to neuronal signalling. *J Inorg Biochem 153*:1–12. https://doi.org/10.1016/j.jinorgbio.2015.09.003

Stephan BC, Hunter S, Harris D, Llewellyn DJ, Siervo M, Matthews FE, Brayne C (2012) The neuropathological profile of mild cognitive impairment (MCI): a systematic review. *Mol Psychiatry 17*(11):1056–1076. https://doi.org/10.1038/mp.2011.147

Stern Y (2012) Cognitive reserve in ageing and Alzheimer's disease. *Lancet Neurol 11*(11):1006–1012. https://doi.org/10.1016/S1474-4422(12)70191-6

Sturchio A, Dwivedi AK, Young CB, Malm T, Marsili L, Sharma JS, Mahajan A, Hill EJ, Andaloussi SE, Poston KL, Manfredsson FP, Schneider LS, Ezzat K, Espay AJ (2021) High cerebrospinal amyloid-β 42 is associated with normal cognition in individuals with brain amyloidosis. EClinicalMedicine 38:100988. https://doi.org/10.1016/j.eclinm.2021.100988

Sultana R, Perluigi M, Butterfield DA (2009) Oxidatively modified proteins in Alzheimer's disease (AD), mild cognitive impairment and animal models of AD: role of Abeta in pathogenesis. *Acta Neuropathol 118*(1):131–150. https://doi.org/10.1007/s00401-009-0517-0

Swomley AM, Butterfield DA (2015) Oxidative stress in Alzheimer disease and mild cognitive impairment: evidence from human data provided by redox proteomics. *Arch Toxicol 89*(10):1669–1680. https://doi.org/10.1007/s00204-015-1556-z

Takeda A (2001) Zinc homeostasis and functions of zinc in the brain. *Biometals 14*(3-4):343–351. https://doi.org/10.1023/a:1012982123386

Tolppanen AM, Williams DM, Lawlor DA (2011) The association of serum ionized calcium and vitamin D with adult cognitive performance. *Epidemiology (Cambridge, Mass.) 22*(1):113–117. https://doi.org/10.1097/EDE.0b013e3181f74683

Tucker KL (2016) Nutrient intake, nutritional status, and cognitive function with aging. *Ann N Y Acad Sci 1367*(1):38–49. https://doi.org/10.1111/nyas.13062

Uttara B, Singh AV, Zamboni P, Mahajan RT (2009) Oxidative stress and neurodegenerative diseases: a review of upstream and downstream antioxidant therapeutic options. *Curr Neuropharmacol 7*(1):65–74. https://doi.org/10.2174/157015909787602823

Van Egroo M, Koshmanova E, Vandewalle G, Jacobs H (2022) Importance of the locus coeruleus-norepinephrine system in sleep-wake regulation: implications for aging and Alzheimer's disease. *Sleep Med Rev 62*:101592. https://doi.org/10.1016/j.smrv.2022.101592

Vassilaki M, Aakre JA, Cha RH, Kremers WK, St Sauver JL, Mielke MM, Geda YE, Machulda MM, Knopman DS, Petersen RC, Roberts RO (2015) Multimorbidity and risk of mild cognitive impairment. *J Am Geriatr Soc 63*(9):1783–1790. https://doi.org/10.1111/jgs.13612

Vega JN, Newhouse PA (2014) Mild cognitive impairment: diagnosis, longitudinal course, and emerging treatments. *Curr Psychiatry Rep 16*(10):490. https://doi.org/10.1007/s11920-014-0490-8

Wang J, Liu Y, Xia Q, Xia Q, Wang B, Yang C, Liang J, Liu X (2020b) Potential roles of telomeres and telomerase in neurodegenerative diseases. *Int J Biol Macromol 163*:1060–1078. https://doi.org/10.1016/j.ijbiomac.2020.07.046

Wang J, Markesbery WR, Lovell MA (2006) Increased oxidative damage in nuclear and mitochondrial DNA in mild cognitive impairment. *J Neurochem 96*(3):825–832. https://doi.org/10.1111/j.1471-4159.2005.03615.x

Wang W, Zhao F, Ma X, Perry G, Zhu X (2020a) Mitochondria dysfunction in the pathogenesis of Alzheimer's disease: recent advances. *Mol Neurodegener 15*(1):30. https://doi.org/10.1186/s13024-020-00376-6

Wang X, Wang W, Li L, Perry G, Lee HG, Zhu X (2014) Oxidative stress and mitochondrial dysfunction in Alzheimer's disease. *Biochim Biophys Acta 1842*(8):1240–1247. https://doi.org/10.1016/j.bbadis.2013.10.015

Ward A, Arrighi HM, Michels S, Cedarbaum JM (2012) Mild cognitive impairment: disparity of incidence and prevalence estimates. *Alzheimer's Dement 8*(1):14–21. https://doi.org/10.1016/j.jalz.2011.01.002

Watt NT, Whitehouse IJ, Hooper NM (2010) The role of zinc in Alzheimer's disease. *Int J Alzheimer's Dis 2011*:971021. https://doi.org/10.4061/2011/971021

Welch G, Tsai LH (2022) Mechanisms of DNA damage-mediated neurotoxicity in neurodegenerative disease. *EMBO Rep 23*(6):e54217. https://doi.org/10.15252/embr.202154217

White KN, Conesa C, Sánchez L, Amini M, Farnaud S, Lorvoralak C, Evans RW (2012) The transfer of iron between ceruloplasmin and transferrins. *Biochim Biophys Acta 1820*(3):411–416. https://doi.org/10.1016/j.bbagen.2011.10.006

World Health Organisation. (2021). *Ageing and health*. https://www.who.int/news-room/fact-sheets/detail/ageing-and-health

World Health Organization. (2020). *Decade of healthy ageing: baseline report*. https://www.who.int/publications/i/item/9789240017900

Yong L, Liu L, Ding T, Yang G, Su H, Wang J, Yang M, Chang J (2021) Evidence of effect of aerobic exercise on cognitive intervention in older adults with mild cognitive impairment. *Front Psychiatry 12*:713671. https://doi.org/10.3389/fpsyt.2021.713671

Zekry D, Herrmann FR, Irminger-Finger I, Ortolan L, Genet C, Vitale AM, Michel JP, Gold G, Krause KH (2010) Telomere length is not predictive of dementia or MCI conversion in the oldest old. *Neurobiol Aging 31*(4):719–720. https://doi.org/10.1016/j.neurobiolaging.2008.05.016

Zhang ZH, Song GL (2021) Roles of selenoproteins in brain function and the potential mechanism of selenium in Alzheimer's disease. *Front Neurosci 15*:646518. https://doi.org/10.3389/fnins.2021.646518

Chapter 3
Nutrigenomics and Trace Elements: Hopes and Hypes for Parkinson's Treatment

Al-Hassan Soliman and Wael Mohamed

Abstract The most common kind of neurological condition leading to mobility disability is Parkinson's disease (PD). This illness is characterized by a wide variety of motor symptoms, such as tremors, rigidity, bradykinesia or akinesia, and postural instability. Nevertheless, motor and non-motor symptoms coexist in a person with PD. In many cases, a clinical diagnosis of PD may be more reliable. Nevertheless, a wide variety of other lab tests may help differentiate PD from other forms of parkinsonism. Treatments for PD include the use of levodopa. It is the gold standard treatment for motor symptoms associated with PD. Levodopa crosses the blood–brain barrier and is converted to dopamine in the substantia nigra's dopamine-receiving neurons (SN). Antioxidants including melatonin, resveratrol, green tea, and lipoic acid have gained widespread attention as potential neuroprotective agents in recent years. As our understanding of nutrigenomics grows, we will be able to pinpoint the role that alterations in diet play in the breakdown of normally functioning systems and the development of pathological diseases. This information might be used to improve neuroprotective mechanisms by dietary changes and the introduction of novel, more beneficial natural compounds. Such mechanisms include increased expression of health-promoting genes and decreased expression of disease-promoting genes after brain damage or other pathologies. There are numerous potential neuroprotective techniques that might include this approach.

A.-H. Soliman
Oral Biology Department, Faculty of Dentistry, Sinai University (SU), Arish, North Sinai Peninsula, Egypt

W. Mohamed (✉)
Department of Basic Medical Sciences, Kulliyyah of Medicine, International Islamic University Malaysia (IIUM), Kuantan, Malaysia

Clinical Pharmacology Department, Menoufia Medical School, Menoufia University, Menoufia, Egypt

W. Mohamed, R. Sandhir (eds.), *Trace Elements in Brain Health and Diseases*, Nutritional Neurosciences, https://doi.org/10.1007/978-981-99-1513-2_3

Graphical Abstract

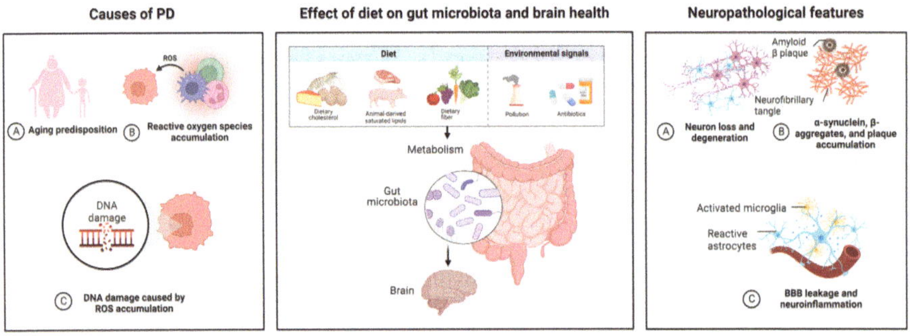

This graphical abstract shows the causes of Parkinson's disease. Etiologies of Parkinson's disease (PD) involve aging and elderly people (A) and accumulation of reactive oxygen species (ROS) in cells (B), which lead to deoxyribonucleic acid (DNA) damage and cell rupture (C). Additionally, diet can affect the gut microbiota and alter the equilibrium between the gut–brain axis that leads to neurodegeneration. Neurodegenerative diseases are characterized by neuron degeneration and loss (A), accumulation of misfolded proteins, α-synucleins, and β-aggregates (B), and, finally, blood–brain barrier leakage that terminates into invasive neuroinflammation (C).

Keywords Nutrigenomics · Parkinson's disease · Brain health · Trace elements · Nutrition

3.1 Introduction

Parkinson's disease (PD) is the most prevalent neurological disorder that causes movement impairment (Balestrino and Schapira 2020; Simon et al. 2020). Numerous motor symptoms, including tremors, stiffness, bradykinesia or akinesia, and postural instability, define this condition. However, Parkinson's disease involves both motor and non-motor symptoms. Parkinson's disease (PD) is an age-related neurodegenerative disorder (Simon et al. 2020; Antony et al. 2013), and it has been observed that aging is the single most significant factor influencing the clinical manifestation and development of PD (Hindle 2010). Age is the most major risk factor for getting PD at the current moment (Antony et al. 2013). PD may often be diagnosed more accurately clinically (Tolosa et al. 2006). However, a vast array of other laboratory tests may distinguish the diagnosis of PD from other Parkinsonisms (Balestrino and Schapira 2020). Dopaminergic neuron loss in the substantia nigra (SN) pars compacta of the brain is a pathological characteristic of Parkinson's disease (Fig. 3.1). Moreover, these synucleinopathies exhibit a build-up of misfolded protein aggregates, which are located as intra-cytoplasmic inclusions known as Lewy bodies (LBs); this accumulation is the pathognomonic feature of the illness (Balestrino and Schapira 2020). LBs are eosinophilic, spherical, intraneuronal

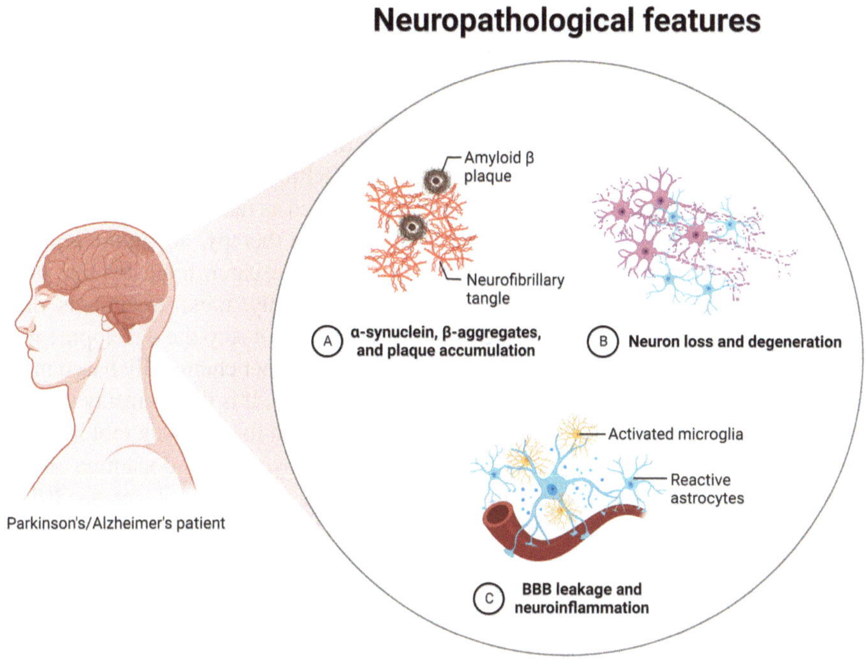

Fig. 3.1 Neuropathological features of Parkinson's disease include remarked accumulation of α-synuclein, beta-amyloid (Aβ)-aggregates, Lewy bodies (LBs), and neurofibrillary tangles. Moreover, this protein accumulation can lead to neuron loss and degeneration. As a result, blood–brain barrier (BBB) leakage and neuroinflammatory responses are activated

inclusions having a hyaline core and a pale periphery. Carriers of mutations in the Parkin gene, which are responsible for the early clinical manifestations of Parkinson's disease, lack the buildup of LBs but exhibit substantial gliosis and neurodegeneration (Antony et al. 2013). As a result of misfolding, synuclein becomes insoluble and forms many amyloid aggregates, which accumulate and create intracellular inclusions. The harmful oligomeric and fibrillar variants of this aggregation process result in mitochondrial impairment, lysosomal and proteasomal malfunction, damage to cellular membranes, and synaptic dysfunction (Antony et al. 2013). Eventually, these characteristics contribute to neural degeneration (Balestrino and Schapira 2020; Samii et al. 2004).

Due in part to mitochondrial deoxyribonucleic acid (mtDNA) damage, neurodegenerative illnesses such as Parkinson's and Alzheimer's diseases are characterized by a high incidence of the inflammatory process (Svilar et al. 2011). It is hypothesized that mtDNA destruction causes or contributes to the etiopathogenesis of various diseases (Fukui and Moraes 2008). Extreme mitochondrial respiratory dysfunction and oxidative phosphorylation result from mtDNA destruction (M. A. Virmani et al. 1995; Nissanka et al. 2018). mtDNA mutations predominantly impact the functioning of complexes I and IV, with complex I related to Parkinson's disease and complex IV to Alzheimer's disease (Hauser and Hastings 2013; Pickrell et al. 2009). It is now evident that PD patients are afflicted by many etiologies that lead to the death of dopaminergic neurons and the development of the illness (Simon et al. 2020). For example, familial history is a risk factor for PD (Balestrino and Schapira 2020). Additionally, environmental pollutants have been related to the genesis and progression of illness (Hindle 2010). Imaging techniques are ineffective in diagnosing Parkinson's disease. Nonetheless, MRI imaging may assist in differential diagnosis by ruling out other brain illnesses, such as malignant and infectious causes (Balestrino and Schapira 2020). As an effective therapy, dopamine replacement treatments are now being considered. However, research indicates that deep brain stimulation improves patient health overall. Sadly, these treatments only address the motor symptoms of the condition. They cannot stop the development of the illness (Balestrino and Schapira 2020). In addition, other chemically based medications, including Levodopa, have been used to treat PD. It is the mainstay therapy for Parkinson's disease and the most effective medication for relieving motor symptoms. Levodopa can cross the BBB, where it is converted to dopamine in the dopamine-receiving neurons of the substantia nigra (SN) (Antony et al. 2013). Levodopa is often administered orally numerous times daily in tablet form. In patients with significant disease development, however, it is administered through duodenal infusion. Numerous studies have found that Levodopa might elicit peripheral adverse effects, including nausea and hypotension. Moreover, it may induce drowsiness, disorientation, and hallucinations (Balestrino and Schapira 2020).

3.2 Trace Elements and Health

The many mechanisms that sustain bodily function and health depend on trace elements (TEs). Age, sex, and nutritional condition are a few factors that play a role in the intricate regulation of TE homeostasis. Negative health effects, such as poor redox homeostasis and genome stability maintenance, may occur if the TE homeostasis is compromised. We sought to comprehend the impacts of brief feeding with appropriate or inadequate levels of four TEs in parallel based on age-related alterations in TEs that have been described in mice well supplied with TEs. Mice that consumed a diet tailored for their age and higher in selenium and zinc were able to halt the aging-related reduction in both TEs. Researchers examined a wide range of endpoints that are indicative of the TE and redox state in detail. They also examined DNA hydroxymethylation and markers that indicate the maintenance of genomic stability (Ekmekcioglu 2001). Age-specific changes in TE concentrations were largely constant and irrespective of dietary intake. Additionally, indicators indicating the redox status were altered and hepatic DNA hydroxymethylation was markedly elevated in the old animals. The TEs' nutritional condition was inconsistently impacted by the lowered supply, with Fe deficiency suffering the most. The disparities between the sexes that were seen in the modifications to the redox status and DNA repair activity may have been influenced by this. The intricacy of factors influencing the TE state and its physiological ramifications is highlighted by our findings overall. When thinking about TE therapies for enhancing general health and promoting convalescence in clinics, it is crucial to consider changes in TE supply, age, and sex.

3.3 Trace Elements in the Aging Process

Given the enormous growth in the number of "baby boomers" and other elderly individuals in the general population, maintaining health with age will be one of the most significant public health concerns of the next millennium. Immune functions (Zn and Se), oxidative stress (Zn, Se, Cu, Fe, and Mn), lean body mass (Cr), bone density (Cu, B, F, and Sr), and insulin sensitivity are influenced by trace element status (Cr). Se, Zn, Mn, and Cr deficiency hazards seem to be the most significant and potentially Cu and B in postmenopausal women. Deficiencies in trace elements lead to altered immunological responses, increased oxidative stress, reduced cognitive processes, glucose intolerance, and osteoporosis. Several supplementation experiments have shown intriguing results, notably in enhancing immunity, lowering lipoperoxidation, avoiding bone density loss, and enhancing glucose tolerance. To avoid the increasing incidence of infections, cardiovascular illnesses, osteoporosis, and diabetes in the elderly, it is crucial to examine these statistics.

3.3.1 Selenium

Antioxidative nutrients for health and well-being include selenium and zinc (Steinbrenner and Klotz 2020). Inadequate dietary intake has been linked to the etiology of age-related illnesses and immunological and cognitive deterioration in aging people. Selenium and zinc are crucial trace minerals. Although both selenium and zinc are frequently promoted as "antioxidants" in mineral supplements, they may have positive effects as parts of enzymes and other proteins that catalyze redox reactions and/or are involved in the maintenance of redox equilibrium (Ekmekcioglu 2001). Insufficiencies in selenium and zinc status are more likely to occur in older people, according to epidemiological statistics; however, these statistical correlations in epidemiological research do not imply a causal relationship. Intervention experiments are rare and have produced erratic and occasionally even negative results. It should be emphasized that the observed shortages in micronutrients may not always be related to poor dietary intake, since the body's ability to absorb certain nutrients or how they interact with other substances may also have an impact on absorption and distribution. Therefore, any dietary supplementation should be used with caution, and anyone considering taking a mineral supplement should consult their doctor first. The role of selenium and zinc in biological antioxidant systems summarizes research on the availability and supplementation of these trace elements to elderly people, and on the potential effects, they may have on aging-related health problems such as type 2 diabetes mellitus and cognitive and brain health decline (Wandt et al. 2021). Free radicals in general and superoxide radical ions are known to be potential cancer-causing agents (Prohaska 1987). Additionally, oxidative stress is connected to the development and progression of cancer (Huang et al. 1999). Copper (Cu) and iron (Fe) both can form reactive oxygen species like hydroxyl radicals (Hsu et al. 2017). These organisms can assault and result in DNA mutations, which are a pathogenic component of cancer. Aging and deficits in micronutrients are frequent causes of cognitive problems. Different essential micronutrients in the diet, such as zinc, copper, iron, and selenium, are important for maintaining and enhancing antioxidant performances or for influencing the complex network of genes (nutrigenomic approach) involved in encoding proteins for biological functions. These nutrients are involved in age-altered biological functions. Deficits or surpluses in trace elements are two aspects related to genomic stability, one of the main causes of cognitive decline. The significance of micronutrients in cognitive impairment relating to genomic stability in an aging population is reported and discussed in this review. In addition, the topic of telomere integrity will be covered concerning aging, cognitive decline, and the micronutrients involved. This chapter will explain how these three factors can be related to one another and why maintaining the homeostasis of micronutrients is crucial for good aging. Genomic instability can be caused by the aging process, micronutrient shortages, and aging (Meramat et al. 2015).

3.3.2 Zinc

Since zinc is most important for preserving metabolic homeostasis in aged subjects, it is expected to have a major impact on antioxidant mechanisms. It preserves the physiological properties of metallothioneins and turns into a crucial component of the antioxidant enzyme superoxide dismutase (which is also known as antioxidants). Zinc also appears to be a powerful antioxidant that aids in cells' fight against oxygen-free radicals (Meramat et al. 2015). More than 1000 proteins, including numerous proteins essential in DNA damage repairs like p53 and copper/zinc superoxide dismutase (CuZnSOD), are found in zinc. These proteins include DNA-binding proteins with zinc fingers. It is possible to hypothesize that inadequate zinc consumption can weaken the antioxidant defense and interfere with the DNA repair system, leaving the cell extremely vulnerable to oxidative DNA damage. The brain is extremely vulnerable to oxidative stress, and impairments in brain function have been linked to a deficit. Experimental animals and humans with zinc deficiencies have impaired cognitive performance (Sandstead et al. 2000; Salgueiro et al. 2002) (Savarino et al. 2001). Elderly people may experience zinc shortage due to intestinal malabsorption, decreased dietary intake of zinc, insufficient chewing, and low socioeconomic level (Ekmekcioglu 2001; Prohaska 1987). Additionally, poor mastication caused by an inadequate diet may be linked to tooth loss (Mocchegiani et al. 2006). The presence of more zinc in older adults' urine suggests that the metabolism of zinc may change as people age (Hill et al. 2005). According to a prior study, 31% of the older patients who were recruited had low zinc levels, which indicates zinc inadequacy (Marcellini et al. 2006). Zn also contributes to the reduction in oxidative stress. It stops the thiol and iron reactions that result in the production of free radicals. In addition to damaging genetic information, the activated oxygen species can deactivate membrane-bound enzymes and cause lipid peroxidation in cell membranes. Additionally, it is a crucial component of the enzymes that repair nucleic acids. In the presence of zinc cofactors, superoxide dismutase reduces superoxide radicals to hydrogen peroxide. A couple of selenium glutathione peroxidases then convert hydrogen peroxide to water. Effective elimination of these superoxide free radicals lowers the risk of cancer, slows the aging process, and maintains the integrity of the membrane (Savarino et al. 2001). Zinc deficiency in rats may also result in oxidative stress and physical disruption of the blood–brain barrier. Therefore, DNA damage brought on by zinc deficiency may be particularly susceptible to and vulnerable in the brain. To test this, a different investigation was conducted utilizing a rat glioma cell line (C6 cells). To do this, they examined how oxidative stress and DNA damage were affected by reduced intracellular zinc levels in rat glioma cells (Ho and Ames 2002).

3.3.3 Copper

When copper levels become too high, it becomes hazardous. Copper is an essential micronutrient. An example of a transition metal is copper. It participates in biological reduction–oxidation (redox) reactions, making it a key cofactor for numerous redox enzymes. Additionally, the metal may become overloaded, which leaves it vulnerable to redox reactions of the Fenton type that result in oxidative cell damage and death. Copper builds up in the body of a person because it is ingested through food and daily piping-in drinking water, which can result in accumulation (Dietrich et al. 2004). Recent research in a sizable population longitudinal research shows the existence of a link between copper consumption and mental impairment in healthy persons who consumed a lot of saturated and transfats. The comparison is between those whose average daily copper intake is 2.75 mg, which causes their rate of mental deterioration to be about 50% higher than that of people whose average daily copper intake is 0.88 mg (Brewer 2015). According to a recent cohort study, high levels of free copper are strongly related to cognitive loss in otherwise healthy elderly people, suggesting that bound copper may not be as important for cognitive function as previously thought (Salustri et al. 2010). This might be because molecules with modest molecular weights and not too big of a molecule can pass through the blood–brain barrier quite easily. Only free copper can therefore enter the brain (Salustri et al. 2010). The Rancho Bernardo Study, a cross-sectional study, found that trace element levels vary by gender. According to the study, women had higher levels of plasma copper, which is adversely related to cognitive function. (Martin et al. 2008). Higher amounts in women taking postmenopausal estrogen may be largely to blame for this. On the other hand, a different study demonstrates that consuming 8 mg of copper once a day did not enhance cognition or slow the progression of AD (Kessler et al. 2008). The fact that copper supplementation does not affect cognitive performance could be attributed to the 12-month supplementation period short duration (Kessler et al. 2008). Even today, copper is crucial for healthy brain growth and development. However, amyloid (A) and neurofibrillary tangles may form and be neurotoxic as a result of copper dyshomeostasis (Kessler et al. 2008). Additionally, prior research indicates that those with diets high in saturated and transfats and deficient in non-hydrogenated unsaturated fats are at an increased risk of developing AD and cognitive impairment (Morris et al. 2003) (Morris et al. 2004). The ingestion of trace amounts of drinking water may exacerbate the neurodegenerative alterations brought on by a hypercholesterolemic diet, according to animal research. Larger community studies have found a link between high copper intake and a noticeably faster rate of cognitive deterioration, although this was only true for those who had eaten a diet heavy in saturated and transfats. Additionally, dietary copper may impede the removal of A from the brain and may even encourage accumulation and aggregation, all of which contribute to cognitive impairment (Sparks and Schreurs 2003).

3.3.4 Chromium

The so-called glucose tolerance factor, which is made up of two molecules of nicotinic acid and the trace element chromium, has been shown to improve how well insulin works (Beard 2003). The metabolism of lipids and carbohydrates most likely requires chromium. Lack of chromium or the biologically active form of the mineral, the glucose tolerance factor, has been linked to peripheral neuropathy, lipid abnormalities, glucose intolerance, and an increased risk of atherosclerotic disease. The discovery that patients on long-term complete parenteral feeding acquire reduced glucose tolerance, which is corrected by chromium supplementation, has led researchers to hypothesize that chromium plays a function in glucose metabolism (Brown et al. 1986). Age-related declines in tissue chromium levels have been shown (Schroeder et al. 1970), and it appears that dietary chromium consumption is suboptimal for many people, especially the elderly, in Western nations where people consume refined and high-sugar foods (Anderson et al. 1992; Offenbacher et al. 1986). Numerous studies have looked at the impact of chromium supplementation on insulin sensitivity, glucose tolerance, and/or serum lipid profiles. Many of them have shown positive effects on these parameters (Elwood et al. 1982; Lee and Reasner 1994; Potter et al. 1985; Press et al. 1990), whereas others have been unable to do so (Anderson et al. 1983; Offenbacher et al. 1985; Rabinowitz et al. 1983). These disparities may exist, for instance, since not all glucose intolerance is caused by chromium insufficiency, even though many people may be. Numerous research studies also demonstrated how chromium supplementation has positive metabolic effects on aged people. For instance, it was shown in one study that giving older adults 200 lg/d of chromium results in an improvement in their ability to tolerate glucose (Urberg and Zemel 1987), but in contrast to earlier research, more recent studies did not discover such an improvement in glucose tolerance (Urberg and Zemel 1987). In a recent study by Amato et al., healthy older individuals were not shown to benefit from chromium supplementation on insulin improvement in insulin sensitivity, blood lipid levels, or body composition (Wysocka et al. 2011). The variation in the study populations may be the cause of conflicting results in this area. Younger population groupings were excluded from various research studies. Therefore, it is unclear whether chromium has a meaningful therapeutic impact or whether it merely affects elderly people who may have a compromised chromium level (Ekmekcioglu 2001).

3.3.5 *Iron*

With an amount in adults ranging from 2.5 to 4 g, iron is the trace metal found in the body in the highest concentration. The transportation of oxygen (as hemoglobin), storage of oxygen (as myoglobin), and use of oxygen (as cytochromes in the respiratory chain) are the three most significant roles of iron. Additionally, iron is a component of the enzymes catalase and peroxidases, which are essential for the removal of harmful peroxides. Iron can also be found in non-heme enzymes, such as xanthine oxidase. Additionally, recent research indicates that iron affects the performance of two iron regulatory proteins and contributes to NO production. Additionally, there is growing proof that iron controls gene expression. Additionally, cell-mediated immunity likely requires iron, and a lack of it may result in lower lymphocyte counts in the blood and poorer DHT (delayed-type hypersensitivity) skin reaction responses (Ekmekcioglu 2001). Adult men and non-menstruating women should consume 10 mg of potassium daily (Nes et al. 1991). Contrary to zinc depletion, iron deficiency may be more common in poor nations among elderly people of both sexes but is less common in developed nations. For instance, Wright and colleagues measured the nutritional intake and biochemical status of non-institutionalized subjects in Norwich and showed that serum ferritin concentrations increased and the incidence of iron deficiency decreased with advancing age, demonstrating that elderly people who live independently do not have an iron deficiency. Only a small percentage of elderly persons are iron-deficient, according to the results of the Euronut-Seneca Study (Nes et al. 1991) and the third National Health and Nutrition Examination Survey (NHANES III). These researchers also demonstrated that the older population generally consumes enough iron. There is a chance for excessive iron accumulation for a lifetime because daily physiological losses of iron are modest (except in menstrual women), especially when high iron intakes are sustained. As a result, older people rarely experience iron deficiency and, if anything, are at a higher risk for having higher iron storage as they age. Nearly all gastrointestinal blood loss is the primary cause of iron deficiency anemia. In the elderly, non-steroidal anti-inflammatory medicines, cancer, Crohn's disease, and ulcerative colitis are common causes of gastrointestinal bleeding (Goddard et al. 2000). Alterations in dietary habits among the elderly, such as a decrease in meat consumption or an increase in iron inhibitors such as dietary fiber, may potentially affect iron availability. Additionally, iron absorption may be hampered by age-related hypochlorhydria or the use of acid-reducing medicines. Elderly people with low iron levels can experience the characteristics of restless leg syndrome, which is defined by a need to move due to uncomfortable creeping feelings that develop deep inside the legs, in addition to the classic symptoms of anemia (Ekmekcioglu 2001). Iron supplementation has been demonstrated to significantly lessen the symptoms (Ekmekcioglu 2001). Foods high in iron should be offered if an individual's diet is the cause of their iron deficit. If an organic condition, such as gastrointestinal bleeding, is the root of the problem, this organic issue should be resolved, if necessary, along with an iron supplement, typically ferrous sulfate. Since iron can gradually accumulate throughout life, chronic supplementation beyond the dietary requirements of this element should only be carried out in cases of confirmed iron insufficiency. This could be achieved, for instance, by measuring serum ferritin levels, which are good indicators of iron status but may be raised due to cancer, inflammation, or liver illness (Ekmekcioglu 2001). According to recent studies, a more accurate way to assess the body's tissue iron levels is to look at the serum transferrin receptor/log serum ferritin ratio (Punnonen et al. 1997). Uncontrolled chronic iron supplementation, particularly in men with a hereditary predisposition, may cause the body to store excessive levels of iron, which may be hazardous over time. This is due to iron's potent oxidative capabilities, which cause radicals like OH9 to develop. These radicals can cause lipid peroxidation and DNA damage, which, respectively, encourage the growth of cancer and atherosclerosis. For instance, new research has revealed a link between elevated iron reserves and coronary heart disease (Tuomainen et al. 1998).

3.4 Nutrigenomics: Definition and Importance

In his 1864 work, "DerMenschist war erisst" or "A Person Is What They Eat," Ludwig Andreas Feuerbach summarized the profound and lasting effects of nutrition on health. Indeed, Hippocrates remarked, "Let food be your medicine and medicine be your nourishment" (460-377 BCE). Environment and food are the two primary determinants of a person's health or sickness. After the completion of the Human Genome Project (HGP), fresh issues have been raised about whether the food humans consume affects their health. To evaluate the link between nutrition and genes, "Nutrigenomics" research has emerged in response to questions such as whether a person's food affects his or her health on a cellular level owing to metabolic processes (Sales et al. 2014; Virmani et al. 2013). It is the discipline of nutrition that applies molecular methods to seek, access, and analyze the diverse reactions elicited by a certain diet administered to different individuals (Ferguson 2009; Dauncey 2012; Sales et al. 2014). It is the discipline that explains how nutrition may alter gene expression by upregulating or downregulating it (Sales et al. 2014). For example, the interaction between nutrients and a gene activates transcription factors. This relationship improves or inhibits the capacity of transcription factors to bind to regions that lead to RNA polymerase regulation (Fialho et al. 2008). In addition, certain alcoholic beverages, such as wine, may cause the overexpression of the nuclear factor kappa-B (NF-kB), a critical regulator of inflammation and cancer (Sales et al. 2014). Numerous studies have shown that food/diet components may impact genes and their variations, which may have positive health effects (Norheim et al. 2012; Thunders et al. 2013). Following the completion of the HGP, the post-genomic era experienced a research surge in "omics" sciences. These disciplines examine the characterization of many biomolecules, including protein, DNA, RNA, and metabolites (Kussmann and Van Bladeren 2011). With the subsequent emerging of many biological disciplines like proteomics, metabolomics, and transcriptomics (Norheim et al. 2012). Consequently, the original definition of nutrigenomics solely pertained to research on the impact of nutrients/bioactive foods on the gene expression of an individual (Thunders et al. 2013). Recent expansions to this term now include the study of dietary pillars that safeguard the DNA (Daimiel et al. 2012). This emerging area of research tries to understand the effect of dietary components on the genome, transcriptome, proteome, and metabolome (Subbiah 2008; Dauncey 2012; Daimiel et al. 2012). It has been argued that, based on genetic data, nutrigenetics may assist individuals in controlling their genetic susceptibility to health and illness and, therefore, in selecting food that is favorable to their health (Keen 2001; Thunders et al. 2013). Accordingly, the distinction between nutrigenetics and nutrigenomics is that the latter examines the positive or negative interactions between our food and lifestyle and our genomes, resulting in gene expression and regulation (Thunders et al. 2013).

3.5 Trace Elements and Nutrigenomics: The Interplay

As viable candidates for neuroprotection, many research groups began examining the use of antioxidants such as melatonin, resveratrol, green tea, and lipoic acid (Virmani et al. 2013; Wagner et al. 2012). Additionally, Negida and coworkers have elaborated on metabolic substances such as nicotinamide, acetyl-L-carnitine, creatine, and coenzyme Q10 (Negida et al. 2016). They made an unusual effort to propose the antioxidant coenzyme Q10 as a viable and effective therapy for Parkinson's disease. The improvements in nutrigenomics will permit the grasp of molecular and cellular genetics self-regulating pathways, specifically their position in the deterioration of healthy functioning mechanisms and the emergence of pathological disorders. This knowledge might be utilized to increase neuroprotective processes via diet and the usage of new, more advantageous natural chemicals. This might occur via the enhanced production of health-promoting genes and the restriction of disease-promoting genes in brain injury and other disease states. This strategy might be included in a variety of prospective neuroprotective strategies (A. Virmani et al. 2013).

3.6 Trace Elements and Nutrigenomics: Potential Way to PD Management

The interplay between genes and dietary components plays a crucial role in brain cell function, altering gene expression, and regulatory patterns. When these complementary and dynamic components are disrupted, crucial problems in brain health, malfunction, and the development of neurodegenerative illnesses result (Dauncey 2012). Trace elements are recognized to be significant regulators of physiologic and metabolic pathways that are altered throughout the aging process and may thus influence the occurrence of age-related diseases. Optimal consumption is necessary to maintain homeostasis and boost cell defense. Deficiencies are connected to certain diseases. However, the impact of poor trace element intakes throughout life in the onset and severity of age-related chronic illnesses remains unappreciated. In addition, lowering the consumption of many trace elements is particularly challenging for the elderly. This review will utilize selenium as an example to demonstrate how a trace element may affect the aging process and how omics technology can facilitate the research of trace element impacts on the aging process. Future techniques integrating nutrigenomics with durability investigations in humans will contribute to the identification of mechanisms by which trace elements alter the aging process (Méplan 2011). Trace elements may impact metabolic processes that are known to be changed with age, such as oxidative and inflammatory processes, and can assist control the pace at which damage accumulates with time. Consequently, they may be key targets for promoting healthy aging. Trace elements influence processes that are known to be altered with aging, including immune system function (Se, Zn, and Cu) (Méplan 2011; Ferencík and Ebringer 2003; Maggini et al. 2007), oxidative stresses (Se, Zn, and Cu) (Sies 1997), insulin sensitivity (Se and Zn) (Akbaraly et al. 2010), and cognitive function (Se) (M (Gao et al. 2007). Trace elements can influence the aging process and thus the rate of biological aging on three levels: (1) They help regulate oxidative damage and DNA repair efficiency; (2) trace element intake is reduced in older people; and (3) inadequate trace element intake over the long term may increase the likelihood of age-related diseases. Certain trace elements have been found to increase the capacity for repair. In human leukocytes, bleomycin-induced DNA damage is repaired more efficiently when Se, also known as selenomethionine, is available (Méplan 2011). Se is an essential trace element for human health. It plays a crucial role in various metabolically active processes, including the biotransformation of thyroid hormone, antioxidant defense mechanisms, and immunological function. Selenoproteins, which contain Se in the form of the amino acid selenocysteine, carry out the biological tasks of selenium. Human beings include 25 genes that code for selenoproteins (Kryukov et al. 2003). The essential functions of Se in healthy immune response, inflammatory mechanism limitation, cancer prevention, and cardiovascular disease mortality imply that monitoring Se needs may impact the occurrence of chronic illnesses (Bellinger et al. 2009; McGregor 2016). Seo and colleagues found that Se in the form of selenomethionine may activate DNA repair feedback in normal human fibroblasts in vitro and protect cells from DNA damage (Bera et al. 2013). Se as sodium selenate was administered to rats, which affected the plasma levels of haptoglobin, apolipoprotein E, transthyretin, and fibrinogen. These proteins have been associated with a range of age-related illnesses, such as neurological and cardiovascular issues (Méplan 2011). Free radicals and neuroinflammation cascades are responsible for many neurodegenerative diseases, including Parkinson's and Alzheimer's diseases. Curcumin, carotenoids, acetyl-L-carnitine, coenzyme Q10, vitamin D, polyphenols, and other dietary supplements have the potential to tackle many ways in these disorders (A. Virmani et al. 2013).

3.7 Epigenetics: Gene Expression–Nutrition Interface

Gene expression is influenced at the transcriptional, translational, and post-translational levels by nutrition. It is now established that some of these reactions are mediated by epigenetic control. The term epigenetics, which translates to "beyond genetics," refers to the methods that alter gene expression without altering the DNA sequence. These techniques usually entail chemical tagging of chromatin, the notion in which DNA is assembled with histone proteins in the cell nucleus (Dulac 2010). Epigenetic markers induce chromatin remodeling and changes in gene expression. Examples include DNA methylation, which reduces gene activity, and histone changes, such as acetylation, which boost gene activity. Definitions of epigenetics vary greatly, ranging from permanent, heritable alterations to very quick, dynamic, transient changes (Aguilera et al. 2010; Hochberg et al. 2011). The vast majority of neural tissue in adults is not mitotic. Variations vary from steady and heritable to very quick, dynamic, and transient (Meaney and Ferguson-Smith 2010). Therefore, it has been proposed that replication-independent methylation modification and chromatin remodeling may be of more relevance than heritable brain maintenance (Fig. 3.2). Epigenetic control of gene expression is significantly influenced by several environmental variables, including diet, physiological and psychological stress, toxins, and viruses (Meaney and Ferguson-Smith 2010; Zhang and Meaney 2010; Zhang and Meaney 2010). Along with infections, this has a substantial effect on the epigenetics governing gene expression. This, along with the realization that epigenetic marks may be transmitted down across generations, shows that the distinction between inherited and environmental risks for the illness is not as distinct as previously thought. Epigenetic mechanisms regulate gene expression in both normal and disordered brain processes, and they are crucial to the operation of a great number of brain activities. They are changeable and reversible, suggesting a method for environmental health and disease control. Nutrition might be used throughout life to promote mental health and minimize the detrimental

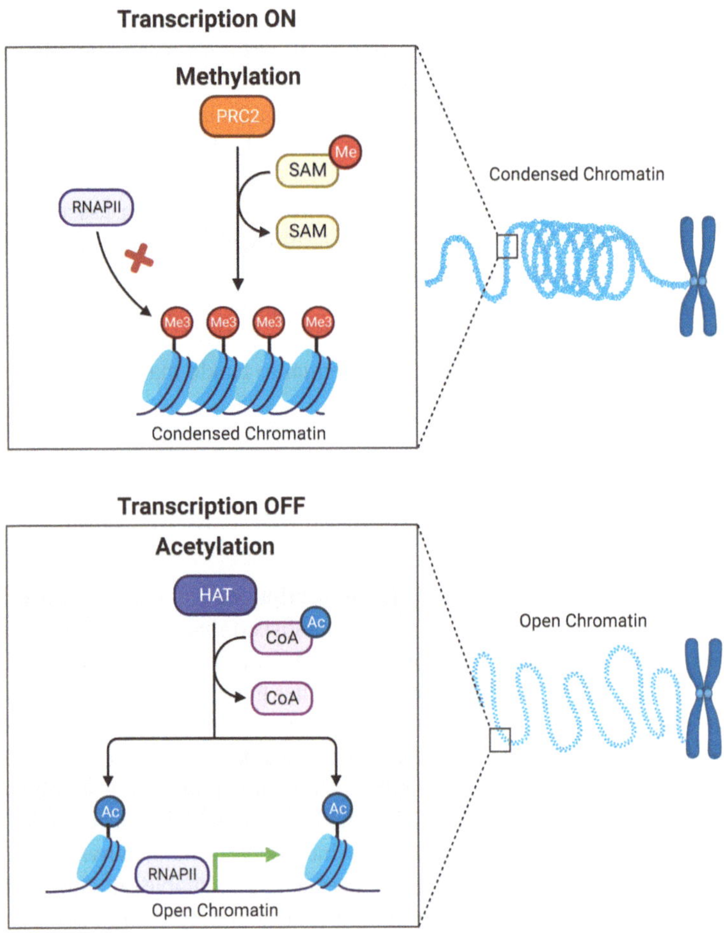

Fig. 3.2 Nutrition can affect gene expression profiles. Certain elements can turn transcription on, and condensed chromatin is found. Gene translation and then transcription occur resulting in gene expression. While other elements can turn it off, it results in open chromatin where gene translation and then transcription does not occur

impacts of early life events. DNA methylation and histone variants are pertinent to the present study due to the importance of research on brain illnesses and the function of nutrition in epigenetic control. These pathways have been associated with cognitive impairment, eating disorders, depression, autism, and schizophrenia. Other epigenetic research has shown a relationship between cognition, eating disorders, depression, autism, and schizophrenia. Other epigenetic processes, including RNA methylation, noncoding microRNA, telomere regulation, and chromosomal position effects, should give a fresh knowledge of the mechanisms behind the relationship between diet and brain health.

3.8 Nutrigenomics and Food–Genome Junction

The emerging subject of nutrigenomics is assisting us in comprehending the reason for some of these variances and promises to enable us to personalize our meals based on our biological profiles. The Human Genome Project's enormous advancements, as well as the recording of single nucleotide polymorphisms (SNPs) at genetic loci with candidate genes and their connection with metabolic abnormalities, have incrementally contributed additional tests to the nutrigenomic framework. According to the notions of ethnopharmacology and phytomedicine, nutrients and herbal extracts are capable of interacting with the genome and producing substantial changes in gene expression. This has led to the commercialization of nutraceuticals and functional foods that may reduce the adverse health impacts of certain genetic profiles, bringing the area to the "food/genome nexus." Incorporating nutrigenomics into preventative medicine may also result in substantial professional improvements (Subbiah 2008). Changes in gene expression influence the impact of diet on the brain (Dauncey 2012). Individual variabilities in food responses are partially attributable to common gene variants including single nucleotides or vast stretches of genomic DNA. Numerous minerals, foods, and diets have been linked to improved brain function (Dauncey 2009; Benton 2010; Gu and Scarmeas 2011; Innis 2011; Morris 2012).

3.9 The Glymphatic System and Brain Health

The glymphatic system, which includes the elimination of brain wastes through perivascular channels regulated by cerebrospinal fluid (CSF), is gaining interest because it provides hitherto unknown insights into neurodegenerative illnesses (Kylkilahti et al. 2021). The glymphatic system is involved in the clearance of metabolic byproducts, such as amyloid, from the central nervous system, and its function is believed to reduce the risk of developing some of the most frequent neurodegenerative illnesses (Kylkilahti et al. 2021; Gallina et al. 2021). CSF is essential for removing harmful compounds from the brain (Kylkilahti et al. 2021). Glymphatic clearance was impaired in the brains of the elderly (Sullan et al. 2018). In general, chemicals generated by brain cells are eliminated from the ISF through a number of mechanisms. The glymphatic system, which includes the elimination of brain wastes through perivascular channels regulated by cerebrospinal fluid (CSF), is gaining interest because it provides hitherto unknown insights into neurodegenerative illnesses (Kylkilahti et al. 2021). The glymphatic system is involved in the clearance of metabolic byproducts, such as amyloid, from the central nervous system, and its function is believed to reduce the risk of developing some of the most frequent neurodegenerative illnesses (Kylkilahti et al. 2021; Gallina et al. 2021). CSF is essential for removing harmful compounds from the brain (Kylkilahti et al. 2021). Glymphatic clearance was impaired in the brains of the elderly (Sullan et al.

2018). In general, chemicals generated by brain cells are ultimately eliminated from the ISF through a number of mechanisms. Prior research found the extracellular tau clearance pathway in the central nervous system, suggesting that glymphatic clearance of extracellular tau is a unique regulatory mechanism whose dysfunction leads to tau accumulation and neurodegeneration (Ishida et al. 2022). The glial–lymphatic ("glymphatic") system, which proposes a mechanism for brain clearing through a peri-vascular CSF flow channel, was investigated. The glymphatic system is indeed implicated in the elimination of amyloid, tau, and a-synuclein. The glymphatic brain clearance process is based on the interchange of cerebrospinal fluid (CSF) and interstitial fluid (ISF), which permits waste to be transferred to CSF and removed from the brain. Because astrocyte end feet are a crucial structural component of the fluid exchange route, the system was named the glia–lymphatic or "glymphatic" system when it was identified in 2012, as astrocyte end feet are a critical structural component of the fluid exchange channel (Iliff et al. 2012). CSF is largely created in the choroid plexus of the 3D lateral ventricles, and it is predominantly transported to the subarachnoid space surrounding the brain by arterial pulsations. The subarachnoid space is contiguous with the periarterial spaces of the pial arteries, from which the CSF penetrates the brain parenchyma and helps solute clearance. The efflux pathways are less well understood. Aquaporin-4 (AQP4) water channels on the end foot of astrocytes that ring the cerebral vasculature are required for the interchange of CSF and ISF. 4,6 Glymphatic dysfunction is associated with alterations in AQP4 expression or polarization (the differential distribution of AQP4 in the end feet vs the rest of the cell). Glymphatic activity may be controlled by several nutritional, behavioral, and physiological modifications in animals, which can be influenced by the human lifestyle (Kylkilahti et al. 2021).

3.10 The Glymphatic–Lymphatic System Interaction

Neurodegeneration is now unequivocally associated with glymphatic–lymphatic system (G-Ls) failure (Nedergaard and Goldman 2020). This linked system in the CNS controls solute transport and immune surveillance. Aquaporin-4 water channels direct CSF circulation from the subarachnoid interiors to the perivascular venues and finally to the interstitial regions (Nedergaard and Goldman 2020; Louveau et al. 2018; Gallina et al. 2021). The cerebrospinal fluid then transports noxious chemicals and immune cells from the brain to the meninges and deep lymph nodes through the venous perivascular and perineural regions. If we consider G-Ls dysfunction to be the most common final pathway to neurodegeneration, we can no longer consider clinical cases in which neurodegeneration manifests as a disease unto itself, but rather pathologies of G-Ls in which noxae and symptoms are centered on a single physiopathological event. G-L injury may alter the morphology and function of CSF spaces by interfering with waste molecule elimination, intra-extracellular water balance, ion homeostasis, and CNS immune response. Aquaporin-4 is also important in ependymal physiology and development (Kahle et al. 2016). These occurrences may result in "G-Ls pathology," a novel category of neurodegenerative disorders based on G-Ls dysfunctions. Thus, Gallina et al. (2021) postulated that neurodegeneration is the ultimate clinical/pathological manifestation of the specific "G-Ls disease" because of several elements classifiable within the framework of cerebral hydrodynamic disturbances.

3.11 Physical Exercise Improves Glymphatic Function

Physical activity promotes brain health by reducing the risk of neurological illnesses and enhancing cognitive performance. Exercise in middle age reduces the chance of acquiring Alzheimer's disease, Parkinson's disease, and dementia decades later (Der Sportwissenschaft 2020). Two studies have shown that voluntary exercise may increase the function of the glymphatic system in rats, which might account for a portion of these positive effects. In terms of glymphatic pathway activation, voluntary jogging substantially increased CSF-ISF exchange and CSF efflux through draining into the deep cervical lymph nodes in comparison with control groups. Exercise resulted in decreased amyloid-b levels, glia cell immunoreactivity, higher AQP4 polarization, and enhanced cognition compared with the sedentary group. Surprisingly, the favorable benefits of exercise on glymphatic function are not evident during acute exercise, indicating that they are not mediated by higher pulse rate and cardiac output. Low noradrenaline levels during sleep or because of pharmaceutical intervention are linked with higher glymphatic function; thus, variations in adrenaline and noradrenaline signaling during exercise might explain the transient reduction in glymphatic inflow during exercise (He et al. 2017; von Holstein-Rathlou et al. 2018).

3.12 Conclusion and Future Perspectives

Intakes and status of trace elements are usually inadequate in the elderly, resulting in nutrition-related illnesses. Considering the essential protective roles of zinc and selenium against oxidative damage and decline of immune functions, as well as the beneficial effect of chromium on glucose tolerance, there is now mounting evidence that appropriate supplementation, restoring marginal trace element status, should reduce the risk of several common age-related degenerative diseases. Institutionalized individuals should also have their antioxidant trace element levels evaluated. Surprisingly, little is known about the actual mineral needs of older individuals. Additionally, additional study is required to examine the efficacy of preventative supplements including many complementary micronutrients. In any case, many scholars agree that greater focus should be placed on the elderly, who represent the ideal target for preventative intervention, as the objective is to keep the aging population in good health.

The use of nutrigenomics in everyday life seems to be the future of nutrition sciences and may aid nutritionists in diagnosing and treating their patients' ailments (Pavlidis et al. 2015). Nonetheless, several issues remain, such as whether nutrigenomics can influence our customized nutrition. Additionally, can it be broadened to influence public health (Pavlidis et al. 2015). Future neuroprotective techniques may include dietary augmentation of neuroprotective pathways and the discovery of more effective natural compounds, i.e., stimulation of health-promoting genes and decrease in the expression of disease-promoting genes (A. Virmani et al. 2013).

References

Aguilera O, Fernández AF, Muñoz A, Fraga MF (2010) Epigenetics and environment: a complex relationship. *J Appl Physiol (Bethesda, Md. : 1985) 109*(1):243–251. https://doi.org/10.1152/japplphysiol.00068.2010

Akbaraly TN, Arnaud J, Rayman MP, Hininger-Favier I, Roussel AM, Berr C, Fontbonne A (2010) Plasma selenium and risk of dysglycemia in an elderly French population: results from the prospective epidemiology of vascular ageing study. *Nutr Metab 7*:1–7. https://doi.org/10.1186/1743-7075-7-21

Anderson RA, Bryden NA, Polansky MM (1992) Dietary chromium intake. Freely chosen diets, institutional diet, and individual foods. *Biol Trace Elem Res 32*:117–121. https://doi.org/10.1007/BF02784595

Anderson RA, Polansky MM, Bryden NA, Roginski EE, Mertz W, Glinsmann W (1983) Chromium supplementation of human subjects: effects on glucose, insulin, and lipid variables. *Metab Clin Exp 32*(9):894–899. https://doi.org/10.1016/0026-0495(83)90203-2

Antony PMA, Diederich NJ, Krüger R, Balling R (2013) The hallmarks of Parkinson's disease. *FEBS J 280*(23):5981–5993. https://doi.org/10.1111/febs.12335

Balestrino R, Schapira AHV (2020) Parkinson disease. *Eur J Neurol 27*(1):27–42. https://doi.org/10.1111/ene.14108

Beard J (2003) Neurocognitive function and trace elements iron deficiency alters brain development and functioning 1,2. *J Nutr 133*(April):1468–1472. https://academic.oup.com/jn/article-abstract/133/5/1468S/4558529

Bellinger FP, Raman AV, Reeves MA, Berry MJ (2009) Regulation and function of selenoproteins in human disease. *Biochem J 422*(1):11–22. https://doi.org/10.1042/BJ20090219

Benton D (2010) The influence of dietary status on the cognitive performance of children. *Mol Nutr Food Res 54*(4):457–470. https://doi.org/10.1002/mnfr.200900158

Bera S, De Rosa V, Rachidi W, Diamond AM (2013) Does a role for selenium in DNA damage repair explain apparent controversies in its use in chemoprevention? *Mutagenesis 28*(2):127–134. https://doi.org/10.1093/mutage/ges064

Brewer GJ (2015) Copper-2 ingestion, plus increased meat eating leading to increased copper absorption, are major factors behind the current epidemic of Alzheimer's Disease. *Nutrients 7*(12):10053–10064. https://doi.org/10.3390/nu7125513

Brown RO, Forloines-Lynn S, Cross RE, Heizer WD (1986) Chromium deficiency after long-term total parenteral nutrition. *Dig Dis Sci 31*(6):661–664. https://doi.org/10.1007/BF01318699

Daimiel L, Vargas T, Ramírez de Molina A (2012) Nutritional genomics for the characterization of the effect of bioactive molecules in lipid metabolism and related pathways. *Electrophoresis 33*(15):2266–2289. https://doi.org/10.1002/elps.201200084

Dauncey MJ (2009) New insights into nutrition and cognitive neuroscience. *Proc Nutr Soc 68*(4):408–415. https://doi.org/10.1017/S0029665109990188

Dauncey MJ (2012) Recent advances in nutrition, genes and brain health. *Proc Nutr Soc 71*(4):581–591. https://doi.org/10.1017/S0029665112000237

Der Sportwissenschaft D. (2020) *Cognitive function and the risk of dementia: the influence of physical fitness and exercise in older adults*. Doctoral Thesis, Cologne, 2020. https://fis.dshs-koeln.de/portal/files/5134977/Diss_Stuckenschneider.pdf

Dietrich AM, Glindemann D, Pizarro F, Gidi V, Olivares M, Araya M, Camper A, Duncan S, Dwyer S, Whelton AJ, Younos T, Subramanian S, Burlingame GA, Khiari D, Edwards M (2004) Health and aesthetic impacts of copper corrosion on drinking water. *Water Sci Technol: A Journal of the International Association on Water Pollution Research 49*(2):55–62

Dulac C (2010) Brain function and chromatin plasticity. *Nature 465*(7299):728–735. https://doi.org/10.1038/nature09231

Ekmekcioglu C (2001) The role of trace elements for the health of elderly individuals. *Die Nahrung 45*(5):309–316. https://doi.org/10.1002/1521-3803(20011001)45:5<309::AID-FOOD309>3.0.CO;2-0

Elwood JC, Nash DT, Streeten DH (1982) Effect of high-chromium brewer's yeast on human serum lipids. *J Am Coll Nutr 1*(3):263–274. https://doi.org/10.1080/07315724.1982.10718995

Ferencík M, Ebringer L (2003) Modulatory effects of selenium and zinc on the immune system. *Folia Microbiol 48*(3):417–426. https://doi.org/10.1007/BF02931378

Ferguson LR (2009) Nutrigenomics approaches to functional foods. *J Am Diet Assoc 109*(3):452–458. https://doi.org/10.1016/j.jada.2008.11.024

Fialho E, Moreno FS, Ong TP (2008) Nutrição no pós-genoma: fundamentos e aplicações de ferramentas ômicas. *Revista de Nutrição 21*(6):757–766. https://doi.org/10.1590/s1415-52732008000600014

Fukui H, Moraes CT (2008) The mitochondrial impairment, oxidative stress and neurodegeneration connection: reality or just an attractive hypothesis? *Trends Neurosci 31*(5):251–256. https://doi.org/10.1016/j.tins.2008.02.008

Gallina P, Nicoletti C, Scollato A, Lolli F (2021) The "Glymphatic-Lymphatic System Pathology" and a new categorization of neurodegenerative disorders. *Front Neurosci 15*:669681. https://doi.org/10.3389/fnins.2021.669681

Gao S, Jin Y, Hall KS, Liang C, Unverzagt FW, Ji R, Murrell JR, Cao J, Shen J, Ma F, Matesan J, Ying B, Cheng Y, Bian J, Li P, Hendrie HC (2007) Selenium level and cognitive function in rural elderly Chinese. *Am J Epidemiol 165*(8):955–965. https://doi.org/10.1093/aje/kwk073

Goddard AF, McIntyre AS, Scott BB (2000) Guidelines for the management of iron deficiency anaemia. British Society of Gastroenterology. *Gut 46 Suppl 3*(Suppl 4):IV1–IV5. https://doi.org/10.1136/gut.46.suppl_4.iv1

Gu Y, Scarmeas N (2011) Dietary patterns in Alzheimer's disease and cognitive aging. *Curr Alzheimer Res 8*(5):510–519. https://doi.org/10.2174/156720511796391836

Hauser DN, Hastings TG (2013) Mitochondrial dysfunction and oxidative stress in Parkinson's disease and monogenic parkinsonism. *Neurobiol Dis 51*:35–42. https://doi.org/10.1016/j.nbd.2012.10.011

He X-F, Liu D-X, Zhang Q, Liang F-Y, Dai G-Y, Zeng J-S, Pei Z, Xu G-Q, Lan Y (2017) Voluntary exercise promotes glymphatic clearance of amyloid beta and reduces the activation of astrocytes and microglia in aged mice. *Front Mol Neurosci 10*:144. https://doi.org/10.3389/fnmol.2017.00144

Hill T, Meunier N, Andriollo-Sanchez M, Ciarapica D, Hininger-Favier I, Polito A, O'Connor JM, Coudray C, Cashman KD (2005) The relationship between the zinc nutritive status and biochemical markers of bone turnover in older European adults: the ZENITH study. *Eur J Clin Nutr 59(Suppl 2)*:S73-8. https://doi.org/10.1038/sj.ejcn.1602303

Hindle JV (2010) Ageing, neurodegeneration and Parkinson's disease. *Age Ageing 39*(2):156–161. https://doi.org/10.1093/ageing/afp223

Ho E, Ames BN (2002) Low intracellular zinc induces oxidative DNA damage, disrupts p53, NFkappa B, and AP1 DNA binding, and affects DNA repair in a rat glioma cell line. *Proc Natl Acad Sci U S A 99*(26):16770–16775. https://doi.org/10.1073/pnas.222679399

Hochberg Z, Feil R, Constancia M, Fraga M, Junien C, Carel J-C, Boileau P, Le Bouc Y, Deal CL, Lillycrop K, Scharfmann R, Sheppard A, Skinner M, Szyf M, Waterland RA, Waxman DJ, Whitelaw E, Ong K, Albertsson-Wikland K (2011) Child health, developmental plasticity, and epigenetic programming. *Endocr Rev 32*(2):159–224. https://doi.org/10.1210/er.2009-0039

von Holstein-Rathlou S, Petersen NC, Nedergaard M (2018) Voluntary running enhances glymphatic influx in awake behaving, young mice. *Neurosci Lett 662*:253–258. https://doi.org/10.1016/j.neulet.2017.10.035

Hsu Y-L, Hung J-Y, Lee Y-L, Chen F-W, Chang K-F, Chang W-A, Tsai Y-M, Chong I-W, Kuo P-L (2017) Identification of novel gene expression signature in lung adenocarcinoma by using next-generation sequencing data and bioinformatics analysis. *Oncotarget 8*(62):104831–104854. https://doi.org/10.18632/oncotarget.21022

Huang YL, Sheu JY, Lin TH (1999) Association between oxidative stress and changes of trace elements in patients with breast cancer. *Clin Biochem 32*(2):131–136. https://doi.org/10.1016/s0009-9120(98)00096-4

Iliff JJ, Wang M, Liao Y, Plogg BA, Peng W, Gundersen GA, Benveniste H, Vates GE, Deane R, Goldman SA, Nagelhus EA, Nedergaard M (2012) A paravascular pathway facilitates CSF flow through the brain parenchyma and the clearance of interstitial solutes, including amyloid β. *Sci Transl Med 4*(147):147ra111. https://doi.org/10.1126/scitranslmed.3003748

Innis SM (2011) Metabolic programming of long-term outcomes due to fatty acid nutrition in early life. *Matern Child Nutr 7 Suppl 2*(Suppl 2):112–123. https://doi.org/10.1111/j.1740-8709.2011.00318.x

Ishida K, Yamada K, Nishiyama R, Hashimoto T, Nishida I, Abe Y, Yasui M, Iwatsubo T (2022) Glymphatic system clears extracellular tau and protects from tau aggregation and neurodegeneration. *J Exp Med 219*:3. https://doi.org/10.1084/jem.20211275

Kahle KT, Kulkarni AV, Limbrick DD, Warf BC (2016) Hydrocephalus in children. *The Lancet 387*(10020):788–799. https://doi.org/10.1016/S0140-6736(15)60694-8

Keen CL (2001) Chocolate: food as medicine/medicine as food. *J Am Coll Nutr 20*(5 Suppl):436S–439S.; discussion 440S-442S. https://doi.org/10.1080/07315724.2001.10719181

Kessler H, Bayer TA, Bach D, Schneider-Axmann T, Supprian T, Herrmann W, Haber M, Multhaup G, Falkai P, Pajonk F-G (2008) Intake of copper has no effect on cognition in patients with mild Alzheimer's disease: a pilot phase 2 clinical trial. *J Neural Transm (Vienna, Austria : 1996) 115*(8):1181–1187. https://doi.org/10.1007/s00702-008-0080-1

Kryukov, G. V, Castellano, S., Novoselov, S. V, Lobanov, A. V, Zehtab, O., Guigó, R., & Gladyshev, V. N. (2003). Characterization of mammalian selenoproteomes *Science (New York, N.Y.), 300*(5624), 1439–1443. https://doi.org/10.1126/science.1083516

Kussmann M, Van Bladeren PJ (2011) The extended nutrigenomics—understanding the interplay between the genomes of food, gut microbes, and human host. *Front Genet* 2:21. https://doi.org/10.3389/fgene.2011.00021

Kylkilahti TM, Berends E, Ramos M, Shanbhag NC, Töger J, Markenroth Bloch K, Lundgaard I (2021) Achieving brain clearance and preventing neurodegenerative diseases-A glymphatic perspective. *J Cereb Blood Flow Metab 41*(9):2137–2149. https://doi.org/10.1177/0271678X20982388

Lee NA, Reasner CA (1994) Beneficial effect of chromium supplementation on serum triglyceride levels in NIDDM. *Diabet Care 17*(12):1449–1452. https://doi.org/10.2337/diacare.17.12.1449

Louveau A, Herz J, Alme MN, Salvador AF, Dong MQ, Viar KE, Herod SG, Knopp J, Setliff JC, Lupi AL, Da Mesquita S, Frost EL, Gaultier A, Harris TH, Cao R, Hu S, Lukens JR, Smirnov I, Overall CC et al (2018) CNS lymphatic drainage and neuroinflammation are regulated by meningeal lymphatic vasculature. *Nat Neurosci 21*(10):1380–1391. https://doi.org/10.1038/s41593-018-0227-9

Maggini S, Wintergerst ES, Beveridge S, Hornig DH (2007) Selected vitamins and trace elements support immune function by strengthening epithelial barriers and cellular and humoral immune responses. *Br J Nutr 98(Suppl 1)*:S29–S35. https://doi.org/10.1017/S0007114507832971

Marcellini F, Giuli C, Papa R, Gagliardi C, Dedoussis G, Herbein G, Fulop T, Monti D, Rink L, Jajte J, Mocchegiani E (2006) Zinc status, psychological and nutritional assessment in old people recruited in five European countries: Zincage study. *Biogerontology 7*(5–6):339–345. https://doi.org/10.1007/s10522-006-9048-4

Martin KC, Ho MV, Martin J-ALR, Buchwald SL, Ho MV, Martin KC, Craik J-AL, Manuscript C, Kantrowitz A (2008) 基因的改变NIH Public Access. *Bone 23*(1):1–7. https://www.ncbi.nlm.nih.gov/pmc/articles/PMC3624763/pdf/nihms412728.pdf

McGregor B (2016) The role of selenium in thyroid autoimmunity: a review. *J Restor Med 4*(1):83–92. https://doi.org/10.14200/jrm.2015.4.0102

Meaney MJ, Ferguson-Smith AC (2010) Epigenetic regulation of the neural transcriptome: the meaning of the marks. *Nat Neurosci 13*(11):1313–1318. https://doi.org/10.1038/nn1110-1313

Méplan C (2011) Trace elements and ageing, a genomic perspective using selenium as an example. *J Trace Elem Med Biol 25*(SUPPL. 1):11–16. https://doi.org/10.1016/j.jtemb.2010.10.002

Meramat A, Rajab NF, Shahar S, Sharif R (2015) Cognitive impairment, genomic instability and trace elements. *J Nutr Health Aging 19*(1):48–57. https://doi.org/10.1007/s12603-014-0489-1

Mocchegiani E, Costarelli L, Giacconi R, Cipriano C, Muti E, Tesei S, Malavolta M (2006) Nutrient-gene interaction in ageing and successful ageing. A single nutrient (zinc) and some target genes related to inflammatory/immune response. *Mech Ageing Dev 127*(6):517–525. https://doi.org/10.1016/j.mad.2006.01.010

Morris MC (2012) Nutritional determinants of cognitive aging and dementia. *Proc Nutr Soc 71*(1):1–13. https://doi.org/10.1017/S0029665111003296

Morris MC, Evans DA, Bienias JL, Tangney CC, Bennett DA, Aggarwal N, Schneider J, Wilson RS (2003) Dietary fats and the risk of incident Alzheimer disease. *Arch Neurol 60*(2):194–200. https://doi.org/10.1001/archneur.60.2.194

Morris MC, Evans DA, Bienias JL, Tangney CC, Wilson RS (2004) Dietary fat intake and 6-year cognitive change in an older biracial community population. *Neurology 62*(9):1573–1579. https://doi.org/10.1212/01.wnl.0000123250.82849.b6

Nedergaard M, & Goldman SA (2020) *10.1126@Science.Abb8739 (1)-1. 56*(October), 50–56.

Negida A, Menshawy A, El Ashal G, Elfouly Y, Hani Y, Hegazy Y, El Ghonimy S, Fouda S, Rashad Y (2016) Coenzyme Q10 for patients with Parkinson's disease: a systematic review and meta-analysis. *CNS Neurol Disord Drug Targets 15*(1):45–53. https://doi.org/10.2174/1871527314666150821103306

Nes M, van Staveren WA, Zajkás G, Inelmen EM, Moreiras-Varela O (1991) Validity of the dietary history method in elderly subjects. Euronut SENECA investigators. *Eur J Clin Nutr 45 Suppl 3*:97–104. http://europepmc.org/abstract/MED/1809573

Nissanka N, Bacman SR, Plastini MJ, Moraes CT (2018) The mitochondrial DNA polymerase gamma degrades linear DNA fragments precluding the formation of deletions. *Nat Commun 9*(1):2491. https://doi.org/10.1038/s41467-018-04895-1

Norheim F, Gjelstad IMF, Hjorth M, Vinknes KJ, Langleite TM, Holen T, Jensen J, Dalen KT, Karlsen AS, Kielland A, Rustan AC, Drevon CA (2012) Molecular nutrition research-The modern way of performing nutritional science. *Nutrients 4*(12):1898–1944. https://doi.org/10.3390/nu4121898

Offenbacher EG, Rinko CJ, Pi-Sunyer FX (1985) The effects of inorganic chromium and brewer's yeast on glucose tolerance, plasma lipids, and plasma chromium in elderly subjects. *Am J Clin Nutr 42*(3):454–461. https://doi.org/10.1093/ajcn/42.3.454

Offenbacher EG, Spencer H, Dowling HJ, Pi-Sunyer FX (1986) Metabolic chromium balances in men. *Am J Clin Nutr 44*(1):77–82. https://doi.org/10.1093/ajcn/44.1.77

Pavlidis C, Patrinos GP, Katsila T (2015) Nutrigenomics: a controversy. *Appl Transl Genom 4*:50–53. https://doi.org/10.1016/j.atg.2015.02.003

Pickrell AM, Fukui H, Moraes CT (2009) The role of cytochrome c oxidase deficiency in ROS and amyloid plaque formation. *J Bioenerg Biomembr 41*(5):453–456. https://doi.org/10.1007/s10863-009-9245-3

Potter JF, Levin P, Anderson RA, Freiberg JM, Andres R, Elahi D (1985) Glucose metabolism in glucose-intolerant older people during chromium supplementation. *Metab Clin Exp 34*(3):199–204. https://doi.org/10.1016/0026-0495(85)90001-0

Press RI, Geller J, Evans GW (1990) The effect of chromium picolinate on serum cholesterol and apolipoprotein fractions in human subjects. *West J Med 152*(1):41–45

Prohaska JR (1987) Functions of trace elements in brain metabolism. *Physiol Rev 67*(3):858–901. https://doi.org/10.1152/physrev.1987.67.3.858

Punnonen K, Irjala K, Rajamäki A (1997) Serum transferrin receptor and its ratio to serum ferritin in the diagnosis of iron deficiency. *Blood 89*(3):1052–1057

Rabinowitz MB, Gonick HC, Levin SR, Davidson MB (1983) Effects of chromium and yeast supplements on carbohydrate and lipid metabolism in diabetic men. *Diabet Care 6*(4):319–327. https://doi.org/10.2337/diacare.6.4.319

Sales NMR, Pelegrini PB, Goersch MC (2014) Nutrigenomics: definitions and advances of this new science. *J Nutr Metab 2014*:202759. https://doi.org/10.1155/2014/202759

Salgueiro MJ, Zubillaga MB, Lysionek AE, Caro RA, Weill R, Boccio JR (2002) The role of zinc in the growth and development of children. *Nutrition 18*(6):510–519. https://doi.org/10.1016/S0899-9007(01)00812-7

Salustri C, Barbati G, Ghidoni R, Quintiliani L, Ciappina S, Binetti G, Squitti R (2010) Is cognitive function linked to serum free copper levels? A cohort study in a normal population. *Clin Neurophysiol 121*(4):502–507. https://doi.org/10.1016/j.clinph.2009.11.090

Samii A, Nutt JG, Ransom BR (2004) Parkinson's disease. *Lancet (London, England) 363*(9423):1783–1793. https://doi.org/10.1016/S0140-6736(04)16305-8

Sandstead HH, Frederickson CJ, Penland JG (2000) History of zinc as related to brain function. *J Nutr 130*(2S Suppl):496S–502S. https://doi.org/10.1093/jn/130.2.496S

Savarino L, Granchi D, Ciapetti G, Cenni E, Ravaglia G, Forti P, Maioli F, Mattioli R (2001) Serum concentrations of zinc and selenium in elderly people: results in healthy nonagenarians/centenarians. *Exp Gerontol 36*(2):327–339. https://doi.org/10.1016/S0531-5565(00)00218-7

Schroeder HA, Nason AP, Tipton IH (1970) Chromium deficiency as a factor in atherosclerosis. *J Chronic Dis 23*(2):123–142. https://doi.org/10.1016/0021-9681(70)90071-8

Sies H (1997) Oxidative stress: oxidants and antioxidants. *Exp Physiol 82*(2):291–295. https://doi.org/10.1113/expphysiol.1997.sp004024

Simon DK, Tanner CM, Brundin P (2020) Parkinson disease epidemiology, pathology, genetics, and pathophysiology. *Clin Geriatr Med 36*(1):1–12. https://doi.org/10.1016/j.cger.2019.08.002

Sparks DL, Schreurs BG (2003) Trace amounts of copper in water induce β-amyloid plaques and learning deficits in a rabbit model of Alzheimer's disease. *Proc Natl Acad Sci U S A 100*(19):11065–11069. https://doi.org/10.1073/pnas.1832769100

Steinbrenner H, Klotz L-O (2020) Selenium and zinc: "antioxidants" for healthy aging? *Zeitschrift fur Gerontologie und Geriatrie 53*(4):295–302. https://doi.org/10.1007/s00391-020-01735-0

Subbiah MTR (2008) Understanding the nutrigenomic definitions and concepts at the food-genome junction. *Omics 12*(4):229–235. https://doi.org/10.1089/omi.2008.0033

Sullan MJ, Asken BM, Jaffee MS, DeKosky ST, Bauer RM (2018) Glymphatic system disruption as a mediator of brain trauma and chronic traumatic encephalopathy. *Neurosci Biobehav Rev 84*:316–324. https://doi.org/10.1016/j.neubiorev.2017.08.016

Svilar D, Goellner EM, Almeida KH, Sobol RW (2011) Base excision repair and lesion-dependent subpathways for repair of oxidative DNA damage. *Antioxid Redox Signal 14*(12):2491–2507. https://doi.org/10.1089/ars.2010.3466

Thunders M, Mangai S, Cooper R (2013) Nutrigenetics, nutrigenomics, and the future of dietary advice. *Food Nutr Sci 04*(10):999–1003. https://doi.org/10.4236/fns.2013.410129

Tolosa E, Wenning G, Poewe W (2006) The diagnosis of Parkinson's disease. *Lancet Neurol 5*(1):75–86. https://doi.org/10.1016/S1474-4422(05)70285-4

Tuomainen TP, Punnonen K, Nyyssönen K, Salonen JT (1998) Association between body iron stores and the risk of acute myocardial infarction in men. *Circulation 97*(15):1461–1466. https://doi.org/10.1161/01.CIR.97.15.1461

Urberg M, Zemel MB (1987) Evidence for synergism between chromium and nicotinic acid in the control of glucose tolerance in elderly humans. *Metab Clin Exp 36*(9):896–899. https://doi.org/10.1016/0026-0495(87)90100-4

Virmani A, Pinto L, Binienda Z, Ali S (2013) Food, nutrigenomics, and neurodegeneration—neuroprotection by what you eat! *Mol Neurobiol 48*(2):353–362. https://doi.org/10.1007/s12035-013-8498-3

Virmani MA, Biselli R, Spadoni A, Rossi S, Corsico N, Calvani M, Fattorossi A, De Simone C, Arrigoni-Martelli E (1995) Protective actions of L-carnitine and acetyl-L-carnitine on the neurotoxicity evoked by mitochondrial uncoupling or inhibitors. *Pharmacol Res 32*(6):383–389. https://doi.org/10.1016/s1043-6618(05)80044-1

Wagner AE, Ernst IMA, Birringer M, Sancak Ö, Barella L, Rimbach G (2012) A combination of lipoic acid plus coenzyme Q10 induces PGC1α, a master switch of energy metabolism, improves stress response, and increases cellular glutathione levels in cultured C2C12 skeletal muscle cells. *Oxid Med Cell Longev 2012*:835970. https://doi.org/10.1155/2012/835970

Wandt VK, Winkelbeiner N, Lossow K, Kopp JF, Schwarz M, Alker W, Nicolai MM, Simon L, Dietzel C, Hertel B, Pohl G, Ebert F, Schomburg L, Bornhorst J, Haase H, Kipp AP, Schwerdtle T (2021) Ageing-associated effects of a long-term dietary modulation of four trace elements in mice. *Redox Biol 46*:102083. https://doi.org/10.1016/j.redox.2021.102083

Wysocka E, Cymerys M, Mielcarz G, Bryl W, Dzięgielewska S, Torliński L (2011) The way of serum chromium utilization may contribute to cardiovascular risk factors in centrally obese persons. *Arch Med Sci 7*(2):257–263. https://doi.org/10.5114/aoms.2011.22076

Zhang T-Y, Meaney MJ (2010) Epigenetics and the environmental regulation of the genome and its function. *Annu Rev Psychol 61*(439–466):C1-3. https://doi.org/10.1146/annurev.psych.60.110707.163625

Chapter 4
Putative Role of Trace Elements Deficiency in Psychiatric Disorders Including Depression

Neda Valian

Abstract Trace elements are essential nutrients exist in all organisms in small amounts. They are involved in many physiological processes in the nervous system, but they have two faces; they may be beneficial and/or detrimental. Trace elements are mainly provided by diet, however, some of them exist in environment. Chronic exposure to high levels of some trace elements can lead to toxic effects on all parts of the body, especially the nervous system. Trace elements play important roles in mental health due to the involvement in learning and memory, mood, affective and social behaviors, cognitive processes, etc. Disturbance in their homeostasis is associated with pathogenesis of mental diseases and neuropsychiatric disorders. In this chapter, the role of trace elements in neuropsychiatric disorders including depression, anxiety, and schizophrenia will be reviewed.

Keywords Trace elements · Zinc · Copper · Manganese · Magnesium · Selenium · Neuropsychiatric disorders · Depression · Anxiety · Schizophrenia

4.1 Neuropsychiatric Disorders

Neuropsychiatric disorders like depression, anxiety, schizophrenia, and obsessive-compulsive disorder affect many people worldwide. Genetic and environmental factors such as psychosocial factors, infectious agents, nutrition, and physical features are involved in the etiology of these disorders. Emerging findings have indicated that the environmental chemical exposures are contributed to the pathogenesis of neuropsychiatric disorders. Dysregulation of synaptic integrity, synaptic plasticity, neurogenesis, neurotransmission of monoamines, balance between excitatory (glutamate) and inhibitory (GABA) neurotransmitters are involved in the pathophysiology of several psychiatric disorders. Besides, trace element homeostasis has

---

N. Valian (✉)
Neuroscience Research Center, Shahid Beheshti University of Medical Sciences, Tehran, Iran
e-mail: n.valian@sbmu.ac.ir

W. Mohamed, R. Sandhir (eds.), *Trace Elements in Brain Health and Diseases*, Nutritional Neurosciences, https://doi.org/10.1007/978-981-99-1513-2_4

played a crucial role in the development of these mental and psychiatric disorders (Hollander et al. 2020).

4.2 Depression

Depression is one of the most common psychiatric disorders with high morbidity and mortality. It affects female more than males and changes all aspect of individuals life including, feelings, thoughts, and behaviors. It has become a global problem and it is estimated as the major cause of suicides (World Health Organization 2017; Stys et al. 2012). In 2014, an epidemiological study indicated that the prevalence of depression ranges from $\leq 3\%$ to 22.5% (in Japan and Afghanistan, respectively) (Smith and De Torres 2014). Recently, the prevalence of major depressive disorder has been estimated 1–3% in adults and children, and affects females more than males (Tekin et al. 2021). Depression is characterized by feelings of persistent sadness, irritability, loss of interest or pleasure in hobbies and activities, decreased energy, fatigue, impairment in concentration and making decisions, sleep disturbance and thoughts of death or suicide, and suicide attempts (Kanter et al. 2008).

Although etiology and underlying mechanisms of depression are complex, however, several factors like diet, sleep, and physical activity are associated with its pathophysiological mechanisms (Nakamura et al. 2019). Glutamatergic system plays an important role in the pathophysiology and treatment of depression. Changes in glutamate levels have been indicated in plasma, cerebrospinal fluid (CSF), and different regions of brain in the patients with depression and in suicide victims (Deutschenbaur et al. 2016). Antagonist of glutamate NMDA receptors, like ketamine, has antidepressant effect (Sowa-Kućma et al. 2013). Furthermore, neuroinflammatory processes are associated with the etiology of depression. Several enzymes including manganese superoxide dismutase (MnSOD), myeloperoxidase (MPO), and inducible nitric oxide synthase (iNOS) are involved in neuroinflammatory and oxidative reactions. In the brain of depressed patients, activity of these enzymes is dysregulated. Increased activity of MPO and iNOS in the patients with depression leads to neuroinflammation, oxidative stress, and neurogenesis deterioration in the brain (Gałecki and Talarowska 2018). In additions, many documents have confirmed the role of different trace elements such as zinc, magnesium, iron, calcium, selenium, manganese, and chromium in depression (Etebary et al. 2010; Xu et al. 2020); however, there is a controversy about their exact role.

4.2.1 Zinc

Zinc (Zn) is one of the essential elements, which is tightly bound to the proteins in the blood with the highest level in the brain, mostly in cortex, hippocampus, amygdala, and olfactory bulb (S. Choi et al. 2020). It involves in the synthesis and degradation of lipids, carbohydrates, proteins, and nucleic acids by regulating the activity of several enzymes like metalloproteinase, DNA polymerase, and RNA polymerase. It also plays a role in cell proliferation, differentiation, and cell membrane stabilization (S. Choi et al. 2020; Mehri 2020). Appropriate level of zinc for daily intake is age- and gender-dependent. Children, women, and men are allowed to intake zinc 3–8 mg, 8–9 mg, and 11 mg, respectively, per day. However, zinc requirements are enhanced to 11–13 mg during pregnancy and lactation. About half of the population receive less than half of the Recommended Dietary Allowance (RDA) of zinc. (Russell et al. 2001).

Zinc deficiency can occur in all age groups, nationalities, and sexes (Cope and Levenson 2010). Despite of the crucial roles in several physiological processes of neurons, excessive levels of zinc can result in neuronal death due to stress oxidative induction in different pathological conditions of the brain like stroke, epileptic seizures, hypoglycemia, and traumatic injuries. On the other hand, zinc deficiency can also induce apoptotic cell death in neurons (D. W. Choi and Koh 1998; Sensi et al. 2009).

Both clinical and experimental evidences have shown a relationship between zinc level and mood disorders like depression, and an inverse relationship between serum zinc concentrations and the severity of depression (Cope and Levenson 2010). Animal studies have shown that zinc has an antidepressant effect in mice and rats, and when administered at very low doses together with ineffective doses of imipramine or citalopram, it could enhance their antidepressant activities (Kroczka et al. 2000; Nowak et al. 2003c; Szewczyk et al. 2002; Szewczyk et al. 2018).

Experimental studies have shown that antidepressant activity of zinc is mediated via antagonistic effect on NMDA receptors (Rosa et al. 2003). In glutamatergic neurons, zinc is colocalized with glutamate in the synaptic vesicles; therefore, zinc can be released from glutamatergic terminals and reacts with NMDA, AMPA, and GABA receptors (S. Choi et al. 2020). Zinc administration in rat could increase inhibitory effect of glycine on NMDA receptors (Cichy et al. 2009). Decrease in the potency of zinc to inhibit NMDA receptors has been reported in the hippocampus of suicide victims (Nowak et al. 2003b; Sowa-Kućma et al. 2013). Furthermore, imipramine, an antidepressant drug, could increase inhibitory effect of zinc on NMDA receptors in the cerebral cortex in mice (Szewczyk et al. 2001). These evidences suggest the involvement of zinc/NMDA receptors interaction in psychopathology of suicide. Antidepressant effect of zinc is also related to its effect on serotonin as an important neurotransmitter in the etiology of depression. It has been indicated that the density of 5-HT1A and 5-HT2A serotonin receptors was significantly increased in the hippocampus and frontal cortex of rat following zinc administration, and antidepressant activity of zinc is prevented by inhibitors of serotonin synthesis and antagonists of serotonin receptors (Cichy et al. 2009). High levels of reactive oxygen species (ROS) has been reported in the brain of patients with depression, therefore increased activity of antioxidant enzymes such as superoxide dismutase (SOD) is crucial for the treatment of depressive symptoms. Chronic oral administration of zinc in rat has antidepressant effect mediated by increased glutathione and BDNF levels in the hippocampus and cerebral cortex (Franco et al. 2008). It has been also indicated that zinc deficiency could decrease proliferation and differentiation of hippocampal and cerebellar neurons in rat, and reduce neurogenesis (Adamo et al. 2010; Gower-Winter et al. 2013).

Clinical evidence has indicated that serum level of zinc was significantly lower in patients with severe treatment resistant depression than normal subjects, while its level is normal in mild depression, suggesting the negative correlation between serum zinc and severity of depression (Maes et al. 1994; Maes et al. 1997). This reduction has also found in elderly population (both men and women) (Anbari-Nogyni et al. 2020) and patients with major depressive disorder (MDD) (Al-Fartusie et al. 2019). In a study performed on Pakistan population, zinc level was compared between depressed patients and age- and gender-matched healthy controls. The findings revealed that serum level of zinc was significantly lower than controls, suggesting the association between zinc concentration and depression (Tanvir et al. 2020).

In 2003, a preliminary clinical study demonstrated that zinc supplementation has beneficial effects in antidepressant therapy in patients with major depression (Nowak et al. 2003a). In a pilot double-blind randomized and placebo-controlled study, it has been demonstrated that zinc supplementation (7 mg for 10 weeks) significantly increased serum zinc levels, improved mood and reduced depressive symptoms (Sawada and Yokoi 2010). Improvement of the symptoms of depression has been also reported in elderly after 70 days of zinc supplementation (Afzali et al. 2021). A meta-analysis of randomized clinical trials and observational studies has demonstrated that monotherapy zinc supplementation could significantly reduce depressive symptom, and the highest level of zinc intake is accompanied with a reduction in the risk of depression (Yosaee et al. 2020). On the other hand, increase in zinc level in the serum and different brain regions, like hippocampus, has been indicated following treatment with antidepressants (like citalopram or imipramine) and electroconvulsive shock (Nowak and Schlegel-Zawadzka 1999). A double-blind placebo-controlled study showed that zinc status has an important role in the efficacy of antidepressant therapy. Patients with treatment-resistant depression could respond to imipramine supplemented with 25 mg zinc per day, in comparison to the patients receiving imipramine alone (Siwek et al. 2009).

4.2.2 *Copper*

Copper (Cu) is an essential trace element for several metabolic processes, iron metabolism, synthesis of hormones and neurotransmitters, and oxidation–reduction reactions. However, high concentrations of copper induce oxidative stress by elevation in free radicals, therefore it can damage proteins and nucleic acids leading to neurotoxicity and neurodegeneration (Vetlényi and Rácz 2020). High levels of copper can also impair zinc uptake which involve in the function of serotonergic neurons, so, interfere to management of depression (Hadi et al. 2002). Therefore, copper hemostasis should be regulated precisely and its level must be kept stable, and an appropriate balance between the concentration of zinc and copper is also important.

Atomic emission spectrometry (AES) in healthy control and patients with MDD has been shown higher serum copper and copper/zinc ratio, and lower serum zinc in patients with MDD compared to healthy controls (Al-Fartusie et al. 2019; Liu et al. 2020). Besides, determination of the levels of the N-acetylaspartate, choline, and creatine levels, as an index of brain activity, using proton magnetic resonance spectroscopy (H-MRS) demonstrated that high level of copper was negatively associated with N-acetylaspartate/choline ratio in the brain, especially prefrontal cortex, suggesting that high-serum level of copper has an inhibitory effects on the regions involved in mood and emotions regulation (Liu et al. 2020). After zinc and iron, copper is the most abundant mineral in the body which is provided by diet. It is transported from the blood to the brain through the blood–brain barrier (BBB), and has an important role in biochemical processes in the central nervous system (Tekin et al. 2021).

Copper, similar to zinc, has modulatory effects on glutamate NMDA and AMPA receptors (Huang et al. 2018; Wapnir 1998). Experimental evidence has documented that copper can inhibit NMDA receptors (Trombley and Shepherd 1996; Vlachová et al. 1996), therefore it is involved in the etiology of depression (Słupski et al. 2019). Furthermore, due to critical role in the regulation of noradrenaline and 5-HT release, and also the function of their receptors, copper may affect the concentration and signaling pathways of these neurotransmitters (Ni et al. 2018). High concentration of copper inhibits dopamine β-hydroxylase activity leading to impaired noradrenaline synthesis (Kornhuber et al. 1994).

An appropriate concentration of copper is also important for tyrosine hydroxylase, a critical enzyme for dopamine synthesis, activity and therefore dopamine level in the brain (Morgan and O'dell 1977). Therefore, copper can play an important role in developing depressive disorders. There is a controversy about the relationship between copper level and depression. A systematic review and meta-analysis has indicated that increased copper levels in the blood may be correlated with symptoms of depression, and it may be considered as a diagnostic biomarker for depression (Ni et al. 2018). Enhanced serum level of copper has been found in women experiencing post-partum depression, in comparison to non-depressed women and men (Crayton and Walsh 2007). Evaluating the school children aged 7–18 years indicated lower zinc and zinc/copper ratio, and higher copper serum level in participants suffered from moderate and severe depression. Surprisingly, physical activity has positive effect on zinc concentration and improves depressive symptoms (Alghadir et al. 2016). Conversely, it has been reported that reduced intake level of copper concomitant with low levels of zinc and manganese is associated with depression (Nakamura et al. 2019). A recent meta-analysis has revealed that the intake levels of copper, manganese, and selenium are negatively related to depression (Ding and Zhang 2022).

4.2.3 Manganese

Manganese (Mn) is an essential component of MnSOD which is responsible for scavenging ROS in oxidative stress. Furthermore, it is involved in protein synthesis, neurotransmitter synthesis, and metabolic processes. Both manganese deficiency and intoxication may be associated with neuropsychiatric disorders (Mehri 2020). Enhanced manganese level can induce neurotoxicity via oxidative stress and, therefore, can lead to several neurological and neuropsychiatric disorders (Chen et al. 2015). It is entered to the brain by active transport across the BBB and can accumulate in the basal ganglia (Racette et al. 2021). Excessive exposure to manganese can also cause Parkinsonian-like motor symptoms and cognitive impairments (Dlamini et al. 2020; Racette et al. 2017). Depression is also found in the population with environmental expose to manganese (Racette et al. 2021; Shiue 2015). A cohort study on women during pregnancy and after delivery has demonstrated that elevated manganese level during pregnancy is an important factor for induction of post-partum depression (Guy et al. 2018; McRae et al. 2020). During pregnancy, manganese can mobilize from internal stores to the blood, and this elevation of manganese in the blood play a role in the development of post-partum depression (Henn et al. 2010). Interestingly, decrease in neurotrophic factors including BDNF has been reported in the people exposed to occupational manganese (Mihailović et al. 2000). The role of low-level BDNF in the pathophysiology of depression has been previously confirmed, thus it can be proposed that higher manganese concentration could induce depression by decreased BDNF level. In contrast, some other studies report opposite findings, indicating lower serum levels of manganese, magnesium, and nickel in patients MDD compared to control (Al-Fartusie et al. 2019). In animal model of depression induced by stress, high levels of manganese, iron, and cobalt in the brain could induce protection against depressive symptoms (Xu et al. 2020).

4.2.4 Magnesium

Magnesium (Mg) is an important intracellular cation involved in the activation of many enzymes. It is the fourth most abundant trace element in the body (Wenwen et al. 2019). Magnesium has an antidepressant-like effects, and its deficiency has been demonstrated in depressed patients (Etebary et al. 2010; Szewczyk et al. 2018). Several studies reported potential positive link between hypomagnesaemia and the risk of depression (Cheungpasitporn et al. 2015; Yary et al. 2016). Negative relation is also reported between magnesium intake level and depression (Jacka et al. 2009). An experimental study has indicated depressive symptoms in mice with low-magnesium diet (Singewald et al. 2004). However, in one study, it has been indicated a moderate decrease in magnesium level only in patients with severe, not mild, and major depression (Nechifor 2009).

It has been proposed that post-partum depression is also related to reduced level of magnesium during pregnancy (Etebary et al. 2010). Magnesium is a potent antagonist of NMDA receptors and its deficiency results in over activity of glutamate. In suicide victims, the potency of magnesium to inhibit NMDA receptors is reduced (Sowa-Kućma et al. 2013). In a randomized, double-blind, placebo-controlled trial, magnesium supplementation (500 mg magnesium oxide tablets for 8 weeks) significantly increased the serum level of magnesium in patients with depression with hypomagnesemia and improves depression symptoms (Rajizadeh et al. 2017). It could also be effective in reducing depression symptoms similar to tricyclic antidepressant imipramine, in diabetic patients (Barragán-Rodríguez et al. 2008). Furthermore, supplementation has been also reported to enhance the efficacy of antidepressants in depressed patients with reduced level of magnesium compared to patients with normal magnesium concentration (Ryszewska-Pokraśniewicz et al. 2018).

4.2.5 Selenium

Selenium (Se) is an essential trace element in humans and animals with an antioxidant activity. It is an important component of glutathione peroxidase (GPx), so an appropriate level of selenium can induce protection against free radical oxidation (Roman et al. 2014; Samad et al. 2022). Since oxidative stress is involved in the physiopathology of depression, an optimal serum level of selenium should be provided. The highest concentrations of selenium exist in liver, kidneys, pancreas, skeletal muscles, thyroid gland, and myocardium. Its level is decreased with aging, smoking, and inflammation (Mehri 2020). A cross-sectional study performed on 2009–2014 indicated that the serum levels of zinc, copper, iron, and selenium are inversely related to depressive symptoms in adults suffering from depression in US (Li et al. 2018). It has been shown that selenium could significantly reduce depressive like behaviors in rat by decrease in the level of malondialdehyde and activity of acetylcholinesterase in the hippocampus (Samad et al. 2022), and by antioxidant, anti-inflammatory effects, and serotonergic signaling modulation as well (Casaril et al. 2019). Selenium-containing compounds such as 3-[(4-methoxyphenyl) selanyl]-2-phenylimidazo[1,2-a] pyridine (MPI) have antidepressant activity mediated by BDNF upregulation, reactive oxygen/nitrogen species reduction, and inhibition of lipid peroxidation prefrontal cortex and hippocampus of mice (Domingues et al. 2019).

In a randomized double-blind placebo-controlled clinical trial, the effect of selenium on postpartum depression was evaluated. Pregnant women were supplemented by selenium (100 mg) from the first trimester of pregnancy until delivery. Depressive symptoms were assessed following delivery for 8 weeks. Selenium supplementation could enhance the serum level of selenium at the time of delivery and significantly reduced postpartum depression in comparison to placebo (Mokhber et al. 2011). Although the underlying mechanism of selenium to attenuate postpartum depression is not clear, however, it has been shown that serum concentration of selenium is progressively declined during pregnancy (Mihailović et al. 2000; Zachara et al. 1993). Since selenium is an essential constituent of GPx involving in lipid peroxidation, which has role in etiology of postmortem depression (Mihailović et al. 2000). So, it could be proposed that selenium exerts its antidepressant effect at least in part due to the antioxidant activity.

4.2.6 Iron

Iron (Fe) is a necessary element for the functions of all tissues. Iron deficiency could induce abnormality in different behaviors through influencing on brain regions like hippocampus and striatum. These two regions are involved in many processes such as learning memory, executive functions such as planning, attention, decision making, emotion, and reward, and are sensitive to iron deficiency (Shah et al. 2021). Iron is involved in the activity of several enzymes and regulation of signal transduction of neurotransmitters like serotonin, dopamine, and noradrenaline. Therefore, it can be considered as an important element in the developing of mood disorders including depression (Gupta 2014) and postpartum depression as well (Etebary et al. 2010). Women with iron deficiency are at risk of postpartum depression three time more than the women with normal iron level (Hameed et al. 2022). It has been shown that ferritin level, an index of iron level, is negatively associated with postpartum depression, up to 32 weeks after delivery, therefore it can be considered as a predicting factor for postpartum depression (Albacar et al. 2011).

4.3 Anxiety

Anxiety is the most common psychiatric disorders among elderly population, which often occurs concurrently with depression (Andreescu and Lee 2020). Anxiety is defined as a feeling of worry or fear, motor tension, and hyperarousal for at least 6 months about anything happens in life which can be mild, moderate, or severe. Anxiety disorders disturb the mood and affect quality of life, education, employment, social interactions, and physical health. The most prevalence of anxiety disorders is observed in people aged 25–44 years, and lowest prevalence is related to subjects aged >65 years (Martin 2022; Vos et al. 2016). Different neurotransmitters play roles in the pathophysiology of anxiety, including monoamines especially serotonin and dopamine, glutamate, and GABA (Młyniec et al. 2015). Therefore, any factors involved in neurotransmitters synthesis, signaling pathways and metabolism (such as several enzymes) (Młyniec et al. 2015), and also in oxidative stress (Hovatta et al. 2010) are associated with both pathophysiology and treatment of anxiety disorders. There are many evidences demonstrating the contribution of trace elements such as zinc, copper, magnesium, manganese, and selenium in anxiety, and their involvement in the regulation of neurotransmitters signaling (Dickerson et al. 2020; Shayganfard 2022). Zinc, copper, and manganese are the cofactors for SOD activity. GPx, as an important antioxidant enzyme, is co-factored by selenium (Hovatta et al. 2010).

4.3.1 Zinc

It has been indicted that zinc deficiency could induce anxiety-like behaviors. In one study, zinc deficiency was induced in mice by zinc-deficient diet, containing 40% of daily zinc requirement. Although the animals had normal locomotor activity, but they showed anxiety-like behavior which was attenuated by desipramine and *Hypericum perforatum* extract (Whittle et al. 2009). Zinc deficiency also induced anxiety in rat indicating by reduced grooming and activity in open field, and decreased open arm time in elevated plus maze (Takeda et al. 2007). Low level of zinc reduction has been reported in elderly population which is associated with anxiety (Anbari-Nogyni et al. 2020), and patients with generalized anxiety disorder (Islam et al. 2013). Moreover, anxiolytic-like effect of zinc has also been shown in rodent. Acute and repeated administration of zinc aspartate could decrease anxiety-like behavior in both rat and mice (Abdel-Maksoud et al. 2012; Partyka et al. 2011). Pretreatment of mice with glutamate NMDA receptors could reduce, and with NMDA antagonist could enhance anxiolytic activity of zinc (Abdel-Maksoud et al. 2012). Consistent with these findings, zinc supplementation (30 mg for 70 days) could decrease the anxiety in elderly, both men and women, in a randomized clinical trial (Afzali et al. 2021).

4.3.2 Copper

Acute copper exposure can reduce anxiety in rodent by changes in serotoninergic and dopaminergic systems. It has been indicated that acute exposure to copper increased immunolabeled serotoninergic cell bodies in the dorsal raphe nucleus, and decreased tyrosine hydroxylase-positive cells in ventral tegmental area and substantia nigra pars compacta, and dopaminergic terminals in striatum. Increased the time and total entries into the open arms in the elevated plus maze and enhanced the time spent in the dark box test were observed as an index of reduced anxiety (Abdellatif and Halima 2017). In patients with generalized anxiety disorder, elevated serum level of copper was seen in comparison to control individuals (Islam et al. 2013).

4.3.3 Magnesium

Experimental evidence showed that low-magnesium diet lead to anxiety-like behavior in mice (Singewald et al. 2004), and magnesium administration exerts anxiolytic activity through inhibitory effect on NMDA receptors (Poleszak et al. 2004). However, several clinical studies have reported no significant correlation between magnesium level and anxiety symptoms. In one study, there was no difference between serum magnesium level in patients with patients with generalized anxiety disorder and healthy controls (Islam et al. 2013). In addition, in patients with anxiety (men and women aged 46–49 or 70–74 years), there was no association between magnesium intake level and anxiety (Jacka et al. 2009). In order to evaluate the effect of magnesium supplementation on postpartum anxiety, the women who delivered in last 48 were supplemented with magnesium (320-mg magnesium sulfate) for 8 weeks. Magnesium supplementation was not effective to reduce postpartum anxiety behaviors (Fard et al. 2017). In a systematic review on studies evaluating the role of magnesium in anxiety disorders (published from 2010 up to 2020), no relationship was found between serum level of magnesium and anxiety in anxious patients (Botturi et al. 2020). On the contrary, there is an evidence indicating negative association between magnesium level and the severity of anxiety. In infertile women, the score of anxiety behavior (Hamilton Anxiety Scale) is higher than control group, and lower magnesium level in peritoneal fluid was accompanied with more severe anxiety. Is has been suggested that reduction in magnesium peritoneal level in patients with anxiety may be due to high levels of catecholamines (Garalejić et al. 2010). Consistent with this, it has been previously reported that adrenaline administration could decrease the level of magnesium in the serum of healthy control men (Joborn et al. 1985). Decrease in magnesium level may be also due to the activation of hypothalamic–pituitary–adrenal (HPA) axis under stress and anxiogenic conditions. Activation of HPA axis results in enhanced aldosterone secretion leading to magnesium excretion (Koeppen and Stanton 2009).

4.3.4 Manganese

Both deficiency and high levels of manganese may lead to anxiety-related behaviors. In the population with environmental manganese exposure, anxiety symptoms are reported (Racette et al. 2021). Consistent with this finding, in the patients with generalized anxiety disorder, serum level of copper is higher than controls (Islam et al. 2013). In an animal study, it has been shown that manganese exposure in pregnant female rat during pregnancy and lactation could reduce anxiety-like behaviors in pups (Molina et al. 2011).

4.3.5 Lead

Chronic elevation in the lead level has detrimental effects on CNS and can change the homeostasis of monoaminergic neurotransmission, especially serotonin (M. Leret et al. 2002). Disturbance in monoaminergic systems results in mood disorders and impairment in motor activity (Sansar et al. 2011; Sansar et al. 2012). Experimental evidence has indicated that chronic lead exposure could induce anxiety-like behaviors in rat (Benammi et al. 2014; Sansar et al. 2012). Exposure to lead for 4 weeks in adult mice increased anxiety simultaneous with memory deficits. Elevated lead levels in the brain significantly enhanced oxidative damage in hippocampus, cerebral cortex, midbrain striatum, and cerebellum. Besides, reduced activity of acetylcholinesterase suggested that lead exerts its neurotoxic effects through acetylcholinesterase activity, redox balance (Ferlemi et al. 2014), and disturbance in monoaminergic systems, especially serotonin in the dorsal raphe nucleus (Benammi et al. 2014). Furthermore, perinatal simultaneous exposure to high level of lead and cadmium could induce anxiety-like behaviors. The underlying mechanism of these behaviors is mainly dysregulation of monoaminergic systems, especially dopaminergic and serotoninergic signaling pathways (M. L. Leret et al. 2003). Daily exposure to lead, from juvenile stage to adult, could induce anxiety-like behaviors in Sprague-Dawley rats via dysregulation of the expression and activity of NMDA and AMPA receptors (Wang et al. 2016).

In a systematic review of observational studies, an association was observed between the level of lead and anxiety disorders (Cybulska et al. 2021). Recently, evaluation of the trace elements in pregnant women demonstrated that enhanced lead concentrations resulted in anxiety symptoms (Levin-Schwartz et al. 2022). Anxiety-like behaviors and impairments in short-term memory are also observed in zebrafish after chronic exposure to lead, characterized by decreased exploratory behaviors and enhanced freezing. Biochemical analysis revealed elevation in cortisol and reduction in the levels of serotonin and melatonin as underlying mechanisms of these behavioral disturbances (Bui Thi et al. 2020).

4.3.6 Selenium

Anxiolytic effect of selenium is reported in several preclinical studies. In an animal model of anxiety induced by arsenic acid, selenium treatment decreased anxiety-like behaviors in rat, by decreasing oxidative stress and malondialdehyde reduction in the hippocampus (Samad et al. 2022). Selenium-containing compounds, such as seleno-organic compound 3-[(4-chlorophenyl) selanyl]-1-methyl-1H-indole (CMI), have been shown to attenuate anxiety-like behavior induced by lipopolysaccharide (LPS) in mice via antioxidant and anti-inflammatory effects. It also modulates serotonergic signaling, and administration of serotonin receptors (5-HT1A, 2A/2C and 3) antagonists could prevent anxiolytic effect of compounds containing selenium.

These findings indicate that selenium reduces anxiety through modulating serotonergic system (Casaril et al. 2019). Another selenium-containing compound, (3-[(4-methoxyphenyl) selanyl]-2-phenylimidazo[1,2-a] pyridine (MPI)), has been also shown to decrease anxiety, through restoration of BDNF expression, reduction of reactive oxygen/nitrogen species, and inhibition of lipid peroxidation in mice prefrontal cortex and hippocampus (Domingues et al. 2019).

4.3.7 Cadmium

Cadmium (Cd) is a highly toxic metal widely existed in the environment and generally been taken by food. It has detrimental effects on CNS, and its high level is associated with behavioral disorders. Although, no association has been reported between cadmium blood level and anxiety (Cybulska et al. 2021), however, recently in an epidemiological study on pregnant women, it has been revealed that elevated levels of cadmium and chromium (Cr) could lead to anxiety-like behaviors (Levin-Schwartz et al. 2022).

4.4 Schizophrenia

Schizophrenia is a psychiatric disorder influencing thoughts, emotions, and different behaviors. The onset of disease is usually at post-pubertal age characterized by positive symptoms including delusion, hallucination, catatonic behavior, and disorganized thinking, due to an increase in activity of neurons (Kozyra et al. 2020). Schizophrenic patients also exhibit negative symptoms like loss of motivation, decreased social behaviors, and cognitive impairments (Camacho-Abrego et al. 2021). Nearly, 1% of adult population are suffering from schizophrenia, and half of them do suicide, at least one time, in their lives (Fields 2009).

This neuropsychiatric disorder has multifactorial and complex etiology, and many factors such as genetic, neurological, environmental and sociological factors, smoking and alcohol use, and lack of exercise have influence on its development. Trace elements are important nutrients playing essential roles in different processes, like metabolic pathways, and functions of the nervous system. It has been indicated that trace elements like zinc, copper, manganese, iron, and selenium are associated with development of schizophrenia (Hale and Test Sr 2010; Shayganfard 2022).

4.4.1 *Zinc*

Many evidences confirmed the role of zinc in the etiology and pathophysiology of neuropsychiatric disorders like schizophrenia. However, there is a controversy about the relationship between zinc concentration and the risk of schizophrenia. Although in some studies no association was found between serum zinc level of schizophrenic patients (Kamkar et al. 2020; Vidović et al. 2013; Yanik et al. 2004), in some studies performed on male patients, decreased concentration of zinc has been shown in the patients with schizophrenia in comparison to healthy controls (Rahman et al. 2009; Tokdemir et al. 2003). In addition, in schizophrenic men and women, low level of serum zinc was elevated after treatment by antipsychotic medication like haloperidol and risperidone (Nechifor et al. 2004). Experimental evidences have presented opposite findings. Neonatal ventral hippocampal lesion (NVHL) with ibotenic acid is an animal model of schizophrenia-exhibited molecular and behavioral changes, as seen in schizophrenic patients (Jones et al. 2011). At pre-pubertal age, increased zinc levels in hippocampus and striatum and, after puberty, elevated zinc concentration have been found in different cortices including prefrontal and occipital of NVHL rat model, which was accompanied with increased neuroinflammation and oxidative stress in cortical and subcortical regions (Camacho-Abrego et al. 2021).

4.4.2 *Copper*

Elevated level of copper has been reported in the patients with schizophrenia (Rahman et al. 2009; Tokdemir et al. 2003; Vidović et al. 2013; Yanik et al. 2004). Conversely, it has been shown that the levels of copper, dysbindin-1 (which modulates copper transportation), and copper transporter-1 (CTR1; which transports copper from the bloodstream to brain across the BBB) are decreased in postmortem hippocampus (Schoonover et al. 2021a), substantia nigra (Schoonover et al. 2018; Schoonover and Roberts 2018), and prefrontal cortex (Schoonover et al. 2021b) of schizophrenic patients. Reduction in the factors involved in copper transportation could disturb copper homeostasis, and subsequently impair neuronal functions (Gokhale et al. 2015). Since dysbindin-1 is involved in cognitive functions, dysregulation or reduction in its level may lead to cognitive impairments (Feng et al. 2008; Papaleo et al. 2012). The symptoms of schizophrenia are resulted from impairments in structural and functional white matter projections (dysconnectivity hypothesis of schizophrenia) (Fornito et al. 2013; Karbasforoushan et al. 2015). Several imaging studies demonstrated abnormality in many white matter tracts connecting cortical and subcortical regions of brain (Agartz et al. 2001; Beasley et al. 2009; Cocchi et al. 2014; Foong et al. 2000; Samartzis et al. 2014; Whitford et al. 2014). Abnormality in the regulation of copper homeostasis, including dysbindin-1, has been revealed in white matter of patients suffering from schizophrenia (Schoonover and Roberts 2021).

High concentrations of zinc exist in the hippocampus, the main region of brain involved in synaptic functions and cognitive behaviors. Reduction in the level of zinc could increase copper concentration. Therefore, lower level of zinc in the brain of schizophrenic patients is accompanied with higher copper level, leading to hyperactivity (Tokdemir et al. 2003). Consistent with this, elevated copper/zinc ratio has been reported in schizophrenic patients compared to healthy control (Farzin et al. 2007; Santa Cruz et al. 2020). Zinc deficiency in women during pregnancy may increase the risk of schizophrenia in babies through enhancement of copper concentration (Andrews 1990; Susser and Lin 1992), which is also demonstrated in rats, mice, and monkeys (Wasantwisut 1997). On the contrary, the other studies indicated no change in serum zinc concentration in the patients with schizophrenia (Herrán et al. 2000; Yanik et al. 2004) and proposed that enhanced level of copper is not attributed to zinc (Herrán et al. 2000). However, there are evidences reporting no change in the serum levels of copper in schizophrenic patients (Cao et al. 2019; Nechifor et al. 2004), both before and after treatment with antipsychotic drugs (Nechifor et al. 2004).

4.4.3 Manganese

Reactive oxygen species (ROS) play a crucial role in physiopathology of schizophrenia. MnSOD in the mitochondria, an important component of antioxidant defense system, could rapidly decrease ROS to protect the cells from oxidative damage (Akyol et al. 2005; Shayganfard 2022). Polymorphisms in the encoding gene of Mn-SOD have been reported to associate with the incidence risk of schizophrenia (Akyol et al. 2005). Early evidence conformed the role of manganese deficiency in the development of schizophrenia (Pfeiffer and LaMola 1983). Consistent with these findings, evaluating the trace elements in the serum of schizophrenic patients aged between 18 and 40 years has indicated decreased level of manganese (Cao et al. 2019). There is a controversy regarding the role of manganese in the etiology of schizophrenia. Some studies reported high concentrations of copper in schizophrenic patients which is accompanied by deficiency of both zinc and manganese (Pfeiffer and LaMola 1999). However, some studies have shown no change in the level of manganese in patient with schizophrenia (Rahman et al. 2009). Conversely, in one study in China population, the level of several trace elements was evaluated in schizophrenic patients. The results indicated that higher concentrations of manganese are significantly associated with an elevated risk of schizophrenia (Ma et al. 2020).

4.4.4 *Magnesium*

Controversial findings are existed regarding the association between the serum magnesium level and schizophrenia. Although no significant relationship between magnesium level and schizophrenia has been previously shown (Athanassenas et al. 1983), however, there is an evidence indicating elevated magnesium level in schizophrenic patients, which is decreased by haloperidol (Jabotinsky-Rubin et al. 1993). Furthermore, in schizophrenic patients with a history of suicide attempt, higher magnesium concentration has been found in platelets in comparison to non-suicidal patients (Ruljancic et al. 2013). In contrast, schizophrenic men and women indicated low level of magnesium in erythrocyte, which is elevated after treatment by antipsychotics haloperidol and risperidone, but its serum level was not changed (Nechifor et al. 2004).

4.4.5 *Selenium*

Selenium (Se), similar to zinc and copper, is a necessary cofactor for the activity of antioxidant enzymes such as SOD and GPx (Shayganfard 2022). In one study, no difference was found between serum selenium level in schizophrenic patients compared to healthy controls (Vidović et al. 2013).

4.5 Conclusion

Trace elements are essential factors for many physiological processes. They work as cofactor for enzymes involved in neuroinflammation, antioxidant defense system, synthesis and metabolism of neurotransmitters, and their signaling pathways. Therefore, trace elements are associated with the pathophysiology of neuropsychiatric disorders. There is a controversy about the relationship between the level of trace elements and the risk of neuropsychiatric diseases. However, it has been demonstrated that zinc level is negatively associated with depression and anxiety, and its high concentration is observed in schizophrenic patients. High levels of copper and manganese increase the risk of depression, anxiety, and schizophrenia.

References

Abdellatif A, Halima G (2017) The neuronal basis of copper induced modulation of anxiety state in rat. Acta Histochem 119(1):10–17

Abdel-Maksoud MS, Abdel-hamid MI, Moustafa YM (2012) NMDA/glutamate mechanism of anxiolytic-like action of zinc in vogel conflict test in mice. IOSR J Pharm 2(5):04–12

Adamo AM, Zago MP, Mackenzie GG, Aimo L, Keen CL, Keenan A, Oteiza PI (2010) The role of zinc in the modulation of neuronal proliferation and apoptosis. Neurotox Res 17(1):1–14

Afzali A, Vakili Z, Goli S, Bagheri H, Mirhosseini S, Ebrahimi H (2021) A randomized clinical trial of the effect of zinc supplement on depression and anxiety in the elderly. Open Public Health J 14:1

Agartz I, Andersson JL, Skare S (2001) Abnormal brain white matter in schizophrenia: a diffusion tensor imaging study. Neuroreport 12(10):2251–2254

Akyol O, Yanik M, Elyas H, Namli M, Canatan H, Akin H et al (2005) Association between ala–9Val polymorphism of Mn-SOD gene and schizophrenia. Prog Neuro-Psychopharmacol Biol Psychiatry 29(1):123–131

Albacar G, Sans T, Martín-Santos R, García-Esteve L, Guillamat R, Sanjuan J et al (2011) An association between plasma ferritin concentrations measured 48 h after delivery and postpartum depression. J Affect Disord 131(1–3):136–142

Al-Fartusie FS, Al-Bairmani HK, Al-Garawi ZS, Yousif AH (2019) Evaluation of some trace elements and vitamins in major depressive disorder patients: a case–control study. Biol Trace Elem Res 189(2):412–419

Alghadir AH, Gabr SA, Al-Eisa E (2016) Effects of physical activity on trace elements and depression related biomarkers in children and adolescents. Biol Trace Elem Res 172(2):299–306

Anbari-Nogyni Z, Bidaki R, Madadizadeh F, Sangsefidi ZS, Fallahzadeh H, Karimi-Nazari E, Nadjarzadeh A (2020) Relationship of zinc status with depression and anxiety among elderly population. Clin Nutr ESPEN 37:233–239

Andreescu C, Lee S (2020) Anxiety disorders in the elderly. Anxiety Disord:561–576

Andrews RR (1990) Unification of the findings in schizophrenia by reference to the effects of gestational zinc deficiency. Med Hypotheses 31(2):141–153

Athanassenas G, Papadopoulos E, Kourkoubas A, Tsitourides S, Gabriel J, Hoïdas S, Frangos E (1983) Serum calcium and magnesium levels in chronic schizophrenics. J Clin Psychopharmacol 3(4):212–216

Barragán-Rodríguez L, Rodríguez-Morán M, Guerrero-Romero F (2008) Efficacy and safety of oral magnesium supplementation in the treatment of depression in the elderly with type 2 diabetes: a randomized, equivalent trial. Magnes Res 21(4):218–223

Beasley CL, Dwork AJ, Rosoklija G, Mann JJ, Mancevski B, Jakovski Z et al (2009) Metabolic abnormalities in fronto-striatal-thalamic white matter tracts in schizophrenia. Schizophr Res 109(1–3):159–166

Benammi H, El Hiba O, Romane A, Gamrani H (2014) A blunted anxiolytic like effect of curcumin against acute lead induced anxiety in rat: involvement of serotonin. Acta Histochem 116(5):920–925

Botturi A, Ciappolino V, Delvecchio G, Boscutti A, Viscardi B, Brambilla P (2020) The role and the effect of magnesium in mental disorders: a systematic review. Nutrients 12(6):1661

Bui Thi NH, Nguyen Thi NA, Audira G, Siregar P, Liang S-T, Huang J-C, Hsiao C-D (2020) Chronic exposure to low concentration lead chloride-induced anxiety and loss of aggression and memory in zebrafish. Int J Mol Sci 21(5):1844

Camacho-Abrego I, González-Cano SI, Aguilar-Alonso P, Brambila E, de la Cruz F, Flores G (2021) Changes in nitric oxide, zinc and metallothionein levels in limbic regions at pre-pubertal and post-pubertal ages presented in an animal model of schizophrenia. J Chem Neuroanat 111:101889

Cao B, Yan L, Ma J, Jin M, Park C, Nozari Y et al (2019) Comparison of serum essential trace metals between patients with schizophrenia and healthy controls. J Trace Elem Med Biol 51:79–85

Casaril AM, Domingues M, de Andrade Lourenço D, Birmann PT, Padilha N, Vieira B et al (2019) Depression-and anxiogenic-like behaviors induced by lipopolysaccharide in mice are reversed by a selenium-containing indolyl compound: Behavioral, neurochemical and computational insights involving the serotonergic system. J Psychiatr Res 115:1–12

Chen P, Chakraborty S, Mukhopadhyay S, Lee E, Paoliello MM, Bowman AB, Aschner M (2015) Manganese homeostasis in the nervous system. J Neurochem 134(4):601–610

Cheungpasitporn W, Thongprayoon C, Mao M, Srivali N, Ungprasert P, Varothai N et al (2015) Hypomagnesaemia linked to depression: a systematic review and meta-analysis. Intern Med J 45(4):436–440
Choi DW, Koh JY (1998) Zionc and brain injury. Annu Rev Neurosci 21:347
Choi S, Hong DK, Choi BY, Suh SW (2020) Zinc in the brain: friend or foe? Int J Mol Sci 21(23):8941
Cichy A, Sowa-Kućma M, Legutko B, Pomierny-Chamioło L, Siwek A, Piotrowska A et al (2009) Zinc-induced adaptive changes in NMDA/glutamatergic and serotonergic receptors. Pharmacol Rep 61(6):1184–1191
Cocchi L, Harding IH, Lord A, Pantelis C, Yucel M, Zalesky A (2014) Disruption of structure–function coupling in the schizophrenia connectome. NeuroImage Clin 4:779–787
Cope EC, Levenson CW (2010) Role of zinc in the development and treatment of mood disorders. Curr Opin Clin Nutr Metab Care 13(6):685–689
Crayton JW, Walsh WJ (2007) Elevated serum copper levels in women with a history of post-partum depression. J Trace Elem Med Biol 21(1):17–21
Cybulska AM, Grochans S, Kamińska MS, Bosiacki M, Skonieczna-Żydecka K, Grochans E (2021) Are cadmium and lead levels linked to the development of anxiety and depression?-a systematic review of observational studies. Ecotoxicol Environ Saf 216:112211
Deutschenbaur L, Beck J, Kiyhankhadiv A, Mühlhauser M, Borgwardt S, Walter M et al (2016) Role of calcium, glutamate and NMDA in major depression and therapeutic application. Prog Neuro-Psychopharmacol Biol Psychiatry 64:325–333
Dickerson AS, Wu AC, Liew Z, Weisskopf M (2020) A scoping review of non-occupational exposures to environmental pollutants and adult depression, anxiety, and suicide. Curr Environ Health Rep 7(3):256–271
Ding J, Zhang Y (2022) Associations of dietary copper, selenium, and manganese intake with depression: a meta-analysis of observational studies. Front Nutr 9
Dlamini WW, Nelson G, Nielsen SS, Racette BA (2020) Manganese exposure, parkinsonian signs, and quality of life in South African mine workers. Am J Ind Med 63(1):36–43
Domingues M, Casaril AM, Birmann PT, Bampi SR, de Andrade Lourenço D, Vieira BM et al (2019) Effects of a selanylimidazopyridine on the acute restraint stress-induced depressive-and anxiety-like behaviors and biological changes in mice. Behav Brain Res 366:96–107
Etebary S, Nikseresht S, Sadeghipour HR, Zarrindast MR (2010) Postpartum depression and role of serum trace elements. Iran J Psychiatry 5(2):40
Fard FE, Mirghafourvand M, Mohammad-Alizadeh Charandabi S, Farshbaf-Khalili A, Javadzadeh Y, Asgharian H (2017) Effects of zinc and magnesium supplements on postpartum depression and anxiety: a randomized controlled clinical trial. Women Health 57(9):1115–1128
Farzin D, Mansouri N, Yazdani T, Ebrahimi P, Zarghami M, Azari P, Hosseini SH (2007) Elevated plasma copper/zinc ratios in patients with schizophrenia. J Gorgan Univ Med Sci 9(1):14–19
Feng Y-Q, Zhou Z-Y, He X, Wang H, Guo X-L, Hao C-J et al (2008) Dysbindin deficiency in sandy mice causes reduction of snapin and displays behaviors related to schizophrenia. Schizophr Res 106(2–3):218–228
Ferlemi A-V, Avgoustatos D, Kokkosis AG, Protonotarios V, Constantinou C, Margarity M (2014) Lead-induced effects on learning/memory and fear/anxiety are correlated with disturbances in specific cholinesterase isoform activity and redox imbalance in adult brain. Physiol Behav 131:115–122
Fields RD (2009) The other brain: from dementia to schizophrenia, how new discoveries about the brain are revolutionizing medicine and science. Simon and Schuster
Foong J, Maier M, Clark C, Barker G, Miller D, Ron M (2000) Neuropathological abnormalities of the corpus callosum in schizophrenia: a diffusion tensor imaging study. J Neurol Neurosurg Psychiatry 68(2):242–244
Fornito A, Harrison BJ, Goodby E, Dean A, Ooi C, Nathan PJ et al (2013) Functional dysconnectivity of corticostriatal circuitry as a risk phenotype for psychosis. JAMA Psychiatry 70(11):1143–1151

Franco JL, Posser T, Brocardo PS, Trevisan R, Uliano-Silva M, Gabilan NH et al (2008) Involvement of glutathione, ERK1/2 phosphorylation and BDNF expression in the antidepressant-like effect of zinc in rats. Behav Brain Res 188(2):316–323
Gałecki P, Talarowska M (2018) Inflammatory theory of depression. Psychiatr Pol 52(3):437–447
Garalejić E, Bojović-Jović D, Damjanović A, Arsić B, Pantić I, Turjačanin-Pantelić D, Perović M (2010) Hamilton anxiety scale (HAMA) in infertile women with endometriosis and its correlation with magnesium levels in peritoneal fluid. Psychiatr Danub 22(1):64–67
Gokhale A, Vrailas-Mortimer A, Larimore J, Comstra HS, Zlatic SA, Werner E et al (2015) Neuronal copper homeostasis susceptibility by genetic defects in dysbindin, a schizophrenia susceptibility factor. Hum Mol Genet 24(19):5512–5523
Gower-Winter SD, Corniola RS, Morgan TJ, Levenson CW (2013) Zinc deficiency regulates hippocampal gene expression and impairs neuronal differentiation. Nutr Neurosci 16(4):174–182
Gupta C (2014) Role of iron (Fe) in body. IOSR J Appl Chem 7(11):38–46
Guy M, Accrombessi M, Fievet N, Yovo E, Massougbodji A, Le Bot B et al (2018) Toxics (Pb, Cd) and trace elements (Zn, Cu, Mn) in women during pregnancy and at delivery, South Benin, 2014–2015. Environ Res 167:198–206
Hadi N, Malik A, Azam S, Khan N, Iqbal J (2002) Serotonin–Cu (II)-mediated DNA cleavage: mechanism of copper binding by serotonin. Toxicol In Vitro 16(6):669–674
Hale DP, Test TR Sr (2010) The role of zinc supplementation in the treatment of schizophrenia. J Soc Cogn Psychol 1:12
Hameed S, Naser IA, Al Ghussein MA, Ellulu MS (2022) Is iron deficiency a risk factor for postpartum depression? A case–control study in the Gaza Strip, Palestine. Public Health Nutr 25(6):1631–1638
Henn BC, Ettinger AS, Schwartz J, Téllez-Rojo MM, Lamadrid-Figueroa H, Hernández-Avila M et al (2010) Early postnatal blood manganese levels and children's neurodevelopment. Epidemiology 21(4):433
Herrán A, García-Unzueta MT, Fernández-González MD, Vázquez-Barquero JL, Alvarez C, Amado JA (2000) Higher levels of serum copper in schizophrenic patients treated with depot neuroleptics. Psychiatry Res 94(1):51–58
Hollander JA, Cory-Slechta DA, Jacka FN, Szabo ST, Guilarte TR, Bilbo SD et al (2020) Beyond the looking glass: recent advances in understanding the impact of environmental exposures on neuropsychiatric disease. Neuropsychopharmacology 45(7):1086–1096
Hovatta I, Juhila J, Donner J (2010) Oxidative stress in anxiety and comorbid disorders. Neurosci Res 68(4):261–275
Huang S, Chen L, Bladen C, Stys PK, Zamponi GW (2018) Differential modulation of NMDA and AMPA receptors by cellular prion protein and copper ions. Mol Brain 11(1):1–6
Islam M, Ahmed MU, Mitu SA, Islam MS, Rahman G, Qusar M, Hasnat A (2013) Comparative analysis of serum zinc, copper, manganese, iron, calcium, and magnesium level and complexity of interelement relations in generalized anxiety disorder patients. Biol Trace Elem Res 154(1):21–27
Jabotinsky-Rubin K, Durst R, Levitin L, Moscovich D, Silver H, Lerner J et al (1993) Effects of haloperidol on human plasma magnesium. J Psychiatr Res 27(2):155–159
Jacka FN, Overland S, Stewart R, Tell GS, Bjelland I, Mykletun A (2009) Association between magnesium intake and depression and anxiety in community-dwelling adults: the Hordaland Health Study. Aust N Z J Psychiatry 43(1):45–52
Joborn H, Åkerström G, Ljunghall S (1985) Effects of exogenous catecholamines and exercise on plasma magnesium concentrations. Clin Endocrinol 23(3):219–226
Jones CA, Watson D, Fone K (2011) Animal models of schizophrenia. Br J Pharmacol 164(4):1162–1194
Kamkar MZ, Siamaki Khaboshan F, Marjani A (2020) Comparison of serum zinc levels between patients with schizophrenia and healthy individuals. J Clin Basic Res 4(3):34–39
Kanter JW, Busch AM, Weeks CE, Landes SJ (2008) The nature of clinical depression: symptoms, syndromes, and behavior analysis. Behav Anal 31(1):1–21

Karbasforoushan H, Duffy B, Blackford JU, Woodward ND (2015) Processing speed impairment in schizophrenia is mediated by white matter integrity. Psychol Med 45(1):109–120

Koeppen BM, Stanton BA (2009) Berne & levy physiology, updated edition E-book. Elsevier Health Sciences

Kornhuber J, Lange KW, Kruzik P, Rausch W-D, Gabriel E, Jellinger K, Riederer P (1994) Iron, copper, zinc, magnesium, and calcium in postmortem brain tissue from schizophrenic patients. Biol Psychiatry 36(1):31–34

Kozyra M, Klas J, Szabat M, Samoń-Drzewicka AM, Milanowska J (2020) The relationship between the diet, microelements, macronutrients and vitamins on the schizophrenia–literature analysis. J Educ Health Sport 10(9):369–377

Kroczka B, Zieba A, Dudek D, Pilc A, Nowak G (2000) Zinc exhibits an antidepressant-like effect in the forced swimming test in mice. Pol J Pharmacol 52(5):403–406

Leret M, Garcia-Uceda F, Antonio M (2002) Effects of maternal lead administration on monoaminergic, GABAergic and glutamatergic systems. Brain Res Bull 58(5):469–473

Leret ML, San Millán JA, Antonio MT (2003) Perinatal exposure to lead and cadmium affects anxiety-like behaviour. Toxicology 186(1–2):125–130

Levin-Schwartz Y, Cowell W, Hsu H-HL, Enlow MB, Amarasiriwardena C, Andra SS et al (2022) Metal mixtures are associated with increased anxiety during pregnancy. Environ Res 204:112276

Li Z, Wang W, Xin X, Song X, Zhang D (2018) Association of total zinc, iron, copper and selenium intakes with depression in the US adults. J Affect Disord 228:68–74

Liu X, Zhong S, Li Z, Chen J, Wang Y, Lai S et al (2020) Serum copper and zinc levels correlate with biochemical metabolite ratios in the prefrontal cortex and lentiform nucleus of patients with major depressive disorder. Prog Neuro-Psychopharmacol Biol Psychiatry 99:109828

Ma J, Yan L, Guo T, Yang S, Liu Y, Xie Q et al (2020) Association between serum essential metal elements and the risk of schizophrenia in China. Sci Rep 10(1):1–11

Maes M, D'haese P, Scharpé S, D'Hondt P, Cosyns P, De Broe M (1994) Hypozincemia in depression. J Affect Disord 31(2):135–140

Maes M, Vandoolaeghe E, Neels H, Demedts P, Wauters A, Meltzer HY et al (1997) Lower serum zinc in major depression is a sensitive marker of treatment resistance and of the immune/inflammatory response in that illness. Biol Psychiatry 42(5):349–358

Martin P (2022) The epidemiology of anxiety disorders: a review. Dialogues Clin Neurosci

McRae N, Bello G, Svensson K, Solano-González M, Wright RJ, Niedzwiecki MM et al (2020) Blood manganese levels during pregnancy and postpartum depression: a cohort study among women in Mexico. Neurotoxicology 76:183–190

Mehri A (2020) Trace elements in human nutrition (II)–an update. Int J Prev Med 11

Mihailović M, Cvetković M, Ljubić A, Kosanović M, Nedeljković S, Jovanović I, Pešut O (2000) Selenium and malondialdehyde content and glutathione peroxidase activity in maternal and umbilical cord blood and amniotic fluid. Biol Trace Elem Res 73(1):47–54

Młyniec K, Gaweł M, Doboszewska U, Starowicz G, Pytka K, Davies CL, Budziszewska B (2015) Essential elements in depression and anxiety. Part II. Pharmacol Rep 67(2):187–194

Mokhber N, Namjoo M, Tara F, Boskabadi H, Rayman MP, Ghayour-Mobarhan M et al (2011) Effect of supplementation with selenium on postpartum depression: a randomized double-blind placebo-controlled trial. J Matern Fetal Neonatal Med 24(1):104–108

Molina RM, Phattanarudee S, Kim J, Thompson K, Wessling-Resnick M, Maher TJ, Brain JD (2011) Ingestion of Mn and Pb by rats during and after pregnancy alters iron metabolism and behavior in offspring. Neurotoxicology 32(4):413–422

Morgan R, O'dell B (1977) Effect of copper deficiency on the concentration of catecholamines and related enzyme activity in the rat brain. J Neurochem 28(1):207–213

Nakamura M, Miura A, Nagahata T, Shibata Y, Okada E, Ojima T (2019) Low zinc, copper, and manganese intake is associated with depression and anxiety symptoms in the Japanese working population: findings from the eating habit and Well-being study. Nutrients 11(4):847

Nechifor M (2009) Magnesium in major depression. Magnes Res 22(3):163–166

Nechifor M, Vaideanu C, Palamaru I, Borza C, Mindreci I (2004) The influence of some antipsychotics on erythrocyte magnesium and plasma magnesium, calcium, copper and zinc in patients with paranoid schizophrenia. J Am Coll Nutr 23(5):549S–551S
Ni M, You Y, Chen J, Zhang L (2018) Copper in depressive disorder: a systematic review and meta-analysis of observational studies. Psychiatry Res 267:506–515
Nowak G, Schlegel-Zawadzka M (1999) Alterations in serum and brain trace element levels after antidepressant treatment. Biol Trace Elem Res 67(1):85–92
Nowak G, Siwek M, Dudek D, Ziêba A, Pilc A (2003a) Effect of zinc supplementation on antidepressant therapy in unipolar depression: a preliminary placebo-controlled study. Pol J Pharmacol 55(6):1143–1148
Nowak G, Szewczyk B, Sadlik K, Piekoszewski W, Trela F, Florek E, Pilc A (2003b) Reduced potency of zinc to interact with NMDA receptors in hippocampal tissue of suicide victims. Pol J Pharmacol 55(3):455–460
Nowak G, Szewczyk B, Wieronska JM, Branski P, Palucha A, Pilc A et al (2003c) Antidepressant-like effects of acute and chronic treatment with zinc in forced swim test and olfactory bulbectomy model in rats. Brain Res Bull 61(2):159–164
Papaleo F, Yang F, Garcia S, Chen J, Lu B, Crawley J, Weinberger D (2012) Dysbindin-1 modulates prefrontal cortical activity and schizophrenia-like behaviors via dopamine/D2 pathways. Mol Psychiatry 17(1):85–98
Partyka A, Jastrzębska-Więsek M, Szewczyk B, Stachowicz K, Sławińska A, Poleszak E et al (2011) Anxiolytic-like activity of zinc in rodent tests. Pharmacol Rep 63(4):1050–1055
Pfeiffer CC, LaMola S (1983) Zinc and manganese in the schizophrenias. J Orthomol Psychiatry 12(3):215–234
Pfeiffer CC, LaMola S (1999) Zinc and manganese in the schizophrenias. J Orthomol Med 14:28–48
Poleszak E, Szewczyk B, Kędzierska E, Wlaź P, Pilc A, Nowak G (2004) Antidepressant-and anxiolytic-like activity of magnesium in mice. Pharmacol Biochem Behav 78(1):7–12
Racette BA, Nelson G, Dlamini WW, Hershey T, Prathibha P, Turner JR et al (2021) Depression and anxiety in a manganese-exposed community. Neurotoxicology 85:222–233
Racette BA, Nielsen SS, Criswell SR, Sheppard L, Seixas N, Warden MN, Checkoway H (2017) Dose-dependent progression of parkinsonism in manganese-exposed welders. Neurology 88(4):344–351
Rahman M, Azad MAK, Hossain M, Qusar M, Bari W, Begum F et al (2009) Zinc, manganese, calcium, copper, and cadmium level in scalp hair samples of schizophrenic patients. Biol Trace Elem Res 127(2):102–108
Rajizadeh A, Mozaffari-Khosravi H, Yassini-Ardakani M, Dehghani A (2017) Effect of magnesium supplementation on depression status in depressed patients with magnesium deficiency: a randomized, double-blind, placebo-controlled trial. Nutrition 35:56–60
Roman M, Jitaru P, Barbante C (2014) Selenium biochemistry and its role for human health. Metallomics 6(1):25–54
Rosa AO, Lin J, Calixto JB, Santos ARS, Rodrigues ALS (2003) Involvement of NMDA receptors and L-arginine-nitric oxide pathway in the antidepressant-like effects of zinc in mice. Behav Brain Res 144(1–2):87–93
Ruljancic N, Mihanovic M, Cepelak I, Bakliza A (2013) Platelet and serum calcium and magnesium concentration in suicidal and non-suicidal schizophrenic patients. Psychiatry Clin Neurosci 67(3):154–159
Russell, R., Beard, J. L., Cousins, R. J., Dunn, J. T., Ferland, G., Hambidge, K., … Stoecker, B. (2001). Dietary reference intakes for vitamin A, vitamin K, arsenic, boron, chromium, copper, iodine, iron, manganese, molybdenum, nickel, silicon, vanadium, and zinc. *A report of the panel on micronutrients, subcommittees on upper reference levels of nutrients and of interpretation and uses of dietary reference intakes, and the standing committee on the scientific evaluation of dietary reference intakes food and nutrition board Institute of medicine*, 797

Ryszewska-Pokraśniewicz B, Mach A, Skalski M, Januszko P, Wawrzyniak ZM, Poleszak E et al (2018) Effects of magnesium supplementation on unipolar depression: a placebo-controlled study and review of the importance of dosing and magnesium status in the therapeutic response. Nutrients 10(8):1014
Samad N, Rao T, Bhatti SA, Imran I (2022) Inhibitory effects of selenium on arsenic-induced anxiety-/depression-like behavior and memory impairment. Biol Trace Elem Res 200(2):689–698
Samartzis L, Dima D, Fusar-Poli P, Kyriakopoulos M (2014) White matter alterations in early stages of schizophrenia: a systematic review of diffusion tensor imaging studies. J Neuroimaging 24(2):101–110
Sansar W, Ahboucha S, Gamrani H (2011) Chronic lead intoxication affects glial and neural systems and induces hypoactivity in adult rat. Acta Histochem 113(6):601–607
Sansar W, Bouyatas MM, Ahboucha S, Gamrani H (2012) Effects of chronic lead intoxication on rat serotoninergic system and anxiety behavior. Acta Histochem 114(1):41–45
Santa Cruz EC, Madrid KC, Arruda MA, Sussulini A (2020) Association between trace elements in serum from bipolar disorder and schizophrenia patients considering treatment effects. J Trace Elem Med Biol 59:126467
Sawada T, Yokoi K (2010) Effect of zinc supplementation on mood states in young women: a pilot study. Eur J Clin Nutr 64(3):331–333
Schoonover KE, Farmer CB, Morgan CJ, Sinha V, Odom L, Roberts RC (2021a) Abnormalities in the copper transporter CTR1 in postmortem hippocampus in schizophrenia: a subregion and laminar analysis. Schizophr Res 228:60–73
Schoonover KE, Kennedy WM, Roberts RC (2021b) Cortical copper transporter expression in schizophrenia: interactions of risk gene dysbindin-1. J Neural Transm 128(5):701–709
Schoonover KE, Queern SL, Lapi SE, Roberts RC (2018) Impaired copper transport in schizophrenia results in a copper-deficient brain state: a new side to the dysbindin story. World J Biol Psychiatry
Schoonover KE, Roberts RC (2018) Isoform and protein region abnormalities of dysbindin and copper transporter proteins in postmortem schizophrenia substantia nigra. bioRxiv:343178
Schoonover KE, Roberts RC (2021) Markers of copper transport in the cingulum bundle in schizophrenia. Schizophr Res 228:124–133
Sensi SL, Paoletti P, Bush AI, Sekler I (2009) Zinc in the physiology and pathology of the CNS. Nat Rev Neurosci 10(11):780–791
Shah HE, Bhawnani N, Ethirajulu A, Alkasabera A, Onyali CB, Anim-Koranteng C, Mostafa JA (2021) Iron deficiency-induced changes in the hippocampus, corpus striatum, and monoamines levels that lead to anxiety, depression, sleep disorders, and psychotic disorders. Cureus 13:9
Shayganfard M (2022) Are essential trace elements effective in modulation of mental disorders? Update and perspectives. Biol Trace Elem Res 200(3):1032–1059
Shiue I (2015) Urinary heavy metals, phthalates and polyaromatic hydrocarbons independent of health events are associated with adult depression: USA NHANES, 2011–2012. Environ Sci Pollut Res 22(21):17095–17103
Singewald N, Sinner C, Hetzenauer A, Sartori SB, Murck H (2004) Magnesium-deficient diet alters depression-and anxiety-related behavior in mice—influence of desipramine and Hypericum perforatum extract. Neuropharmacology 47(8):1189–1197
Siwek M, Dudek D, Paul IA, Sowa-Kućma M, Zięba A, Popik P et al (2009) Zinc supplementation augments efficacy of imipramine in treatment resistant patients: a double blind, placebo-controlled study. J Affect Disord 118(1–3):187–195
Słupski J, Słupska A, Szałach ŁP, Włodarczyk A, Górska N, Szarmach J et al (2019) Role of copper and ketamine in major depressive disorder-an update. Psychiatr Danub 31(suppl 3):520–523
Smith K, De Torres I (2014) A world of depression. Nature 515(181):10.1038
Sowa-Kućma M, Szewczyk B, Sadlik K, Piekoszewski W, Trela F, Opoka W et al (2013) Zinc, magnesium and NMDA receptor alterations in the hippocampus of suicide victims. J Affect Disord 151(3):924–931

Stys PK, You H, Zamponi GW (2012) Copper-dependent regulation of NMDA receptors by cellular prion protein: implications for neurodegenerative disorders. J Physiol 590(6):1357–1368
Susser ES, Lin SP (1992) Schizophrenia after prenatal exposure to the Dutch hunger Winter of 1944-1945. Arch Gen Psychiatry 49(12):983–988
Szewczyk B, Brañski P, Wieroñska JM, Palucha A, Pilc A, Nowak G (2002) Interaction of zinc with antidepressants in the forced swimming test in mice. Pol J Pharmacol 54(6):681–686
Szewczyk B, Kata R, Nowak G (2001) Rise in zinc affinity for the NMDA receptor evoked by chronic imipramine is species-specific. Pol J Pharmacol 53(6):641–646
Szewczyk B, Szopa A, Serefko A, Poleszak E, Nowak G (2018) The role of magnesium and zinc in depression: similarities and differences. Magnes Res 31(3):78–89
Takeda A, Tamano H, Kan F, Itoh H, Oku N (2007) Anxiety-like behavior of young rats after 2-week zinc deprivation. Behav Brain Res 177(1):1–6
Tanvir S, Asif N, Qayyum R, Ijaz A, Hafeez A, Ali S (2020) Trace metal profiling in patients with depression in Pakistani population. JPMA 70:1883
Tekin E, Güneş B, Erbaş O (2021) Depression and copper. J Exp Basic Med Sci 2(2):181–187
Tokdemir M, Polat S, Acik Y, Gursu F, Cikim G, Deniz O (2003) Blood zinc and copper concentrations in criminal and noncriminal schizophrenic men. Arch Androl 49(5):365–368
Trombley PQ, Shepherd GM (1996) Differential modulation by zinc and copper of amino acid receptors from rat olfactory bulb neurons. J Neurophysiol 76(4):2536–2546
Vetlényi E, Rácz G (2020) The physiological function of copper, the etiological role of copper excess and deficiency. Orv Hetil 161(35):1488–1496
Vidović B, Đorđević B, Milovanović S, Škrivanj S, Pavlović Z, Stefanović A, Kotur-Stevuljević J (2013) Selenium, zinc, and copper plasma levels in patients with schizophrenia: relationship with metabolic risk factors. Biol Trace Elem Res 156(1):22–28
Vlachová V, Zemková H, Vyklický L Jr (1996) Copper modulation of NMDA responses in mouse and rat cultured hippocampal neurons. Eur J Neurosci 8(11):2257–2264
Vos T, Allen C, Arora M, Barber RM, Bhutta ZA, Brown A et al (2016) Global, regional, and national incidence, prevalence, and years lived with disability for 310 diseases and injuries, 1990–2015: a systematic analysis for the Global Burden of Disease Study 2015. Lancet 388(10053):1545–1602
Wang T, Guan R-L, Liu M-C, Shen X-F, Chen JY, Zhao M-G, Luo W-J (2016) Lead exposure impairs hippocampus related learning and memory by altering synaptic plasticity and morphology during juvenile period. Mol Neurobiol 53(6):3740–3752
Wapnir RA (1998) Copper absorption and bioavailability. Am J Clin Nutr 67(5):1054S–1060S
Wasantwisut E (1997) Nutrition and development: other micronutrients' effect on growth and cognition. Southeast Asian J Trop Med Public Health 28:78–82
Wenwen X, Jing Y, Yingchao S, Qinglu W (2019) The effect of magnesium deficiency on neurological disorders: a narrative review article. Iran J Public Health 48(3):379
Whitford TJ, Lee SW, Oh JS, de Luis-Garcia R, Savadjiev P, Alvarado JL et al (2014) Localized abnormalities in the cingulum bundle in patients with schizophrenia: a diffusion tensor tractography study. NeuroImage Clin 5:93–99
Whittle N, Lubec G, Singewald N (2009) Zinc deficiency induces enhanced depression-like behaviour and altered limbic activation reversed by antidepressant treatment in mice. Amino Acids 36(1):147–158
World Health Organization (2017) Depression and other common mental disorders: global health estimates. World Health Organization
Xu L, Zhang S, Chen W, Yan L, Chen Y, Wen H et al (2020) Trace elements differences in the depression sensitive and resilient rat models. Biochem Biophys Res Commun 529(2):204–209
Yanik M, Kocyigit A, Tutkun H, Vural H, Herken H (2004) Plasma manganese, selenium, zinc, copper, and iron concentrations in patients with schizophrenia. Biol Trace Elem Res 98(2):109–117
Yary T, Lehto SM, Tolmunen T, Tuomainen T-P, Kauhanen J, Voutilainen S, Ruusunen A (2016) Dietary magnesium intake and the incidence of depression: a 20-year follow-up study. J Affect Disord 193:94–98

Yosaee S, Clark CC, Keshtkaran Z, Ashourpour M, Keshani P, Soltani S (2020) Zinc in depression: from development to treatment: a comparative/dose response meta-analysis of observational studies and randomized controlled trials. General Hospital Psychiatry

Zachara B, Wardak C, Didkowski W, Maciag A, Marchaluk E (1993) Changes in blood selenium and glutathione concentrations and glutathione peroxidase activity in human pregnancy. Gynecol Obstet Investig 35(1):12–17

Chapter 5
Trace Elements and Neurodegenerative Diseases

Lahcen Tamegart, Mjid Oukhrib, Hafida El Ghachi, Abdelali Ben Maloui, Abdelaati El khiat, and Halima Gamrani

Abstract Lead, copper, and aluminum neurotoxicity is associated with numerous alterations including behavioral and neurochemical disruptions. Many studies evaluate the possible neurochemical disruption in the brain structures after acute and chronic Pb, Cu, and Mn-exposures and the possible effect on brain structures. Trace elements are related to neurobehavioral alterations. Using immunohistochemical stainings, many studies compared both acute and chronic exposures to heavy metals. The two models of Pb, Cu, and Mn-exposure showed obvious disruptions of many structures.

Keywords Lead · Copper · Manganese · Alzheimer's disease · Parkinson's disease

L. Tamegart (✉)
Biology and Health Unit, Department of Biology, Faculty of Science, Abdelmalek Essaadi University, Tetouan, Morocco

Neurosciences, Pharmacology and Environment Team, Laboratory of Clinical, Experimental and Environmental Neurosciences, Faculty of Medicine and Pharmacy, Cadi Ayyad University, Marrakech, Morocco
e-mail: l.tamegart@uae.ac.ma

M. Oukhrib · H. El Ghachi · A. B. Maloui · H. Gamrani (✉)
Neurosciences, Pharmacology and Environment Team, Laboratory of Clinical, Experimental and Environmental Neurosciences, Faculty of Medicine and Pharmacy, Cadi Ayyad University, Marrakech, Morocco
e-mail: gamrani@uca.ac.ma

A. El khiat
Neurosciences, Pharmacology and Environment Team, Laboratory of Clinical, Experimental and Environmental Neurosciences, Faculty of Medicine and Pharmacy, Cadi Ayyad University, Marrakech, Morocco

Higher Institute of Nursing Professions and Health Techniques in Ouarzazate Ministry of Health, Ouarzazate , Morocco

W. Mohamed, R. Sandhir (eds.), *Trace Elements in Brain Health and Diseases*, Nutritional Neurosciences, https://doi.org/10.1007/978-981-99-1513-2_5

5.1 Introduction

Neurodegenerative disorders such as Alzheimer's disease (AD), Parkinson's disease (PD), and multiple sclerosis disease (MSD) constitute one of the most important diseases attacking specific regions of the brain leading to atrophy in these regions (Cohen et al. 2020; Ganguly et al. 2018).

The appearance of oxidative stress leads to an imbalance of the antioxidant defense systems, which induces a general disorder of the central nervous system and disrupts their functions leading to diseases such as Alzheimer's disease, Parkinson's disease, amyotrophic lateral sclerosis (Lou Gehrig's disease), and multiple sclerosis. In addition, oxidative stress leads to lung, respiratory, digestive, and renal cancers (Roos et al. 2006; Willis et al. 2010).

Therapeutic solutions have been pointed out, but these have not shown satisfactory resolution due to low efficacy. Nowadays, herbal treatments are coming back to the fore because of the effectiveness of drugs, such as antioxidants, polyphenols, and especially flavonoids, which are powerful antioxidants that can inhibit the increase in free radical levels (Zerrouki 2017).

On the other hand, non-essential metals, which have no known biological role in the body, are elements so harmful to vital organs, namely the brain, liver, and kidneys. They have the ability to induce neurotoxicity, hepatotoxicity, and nephrotoxicity, even at low doses. These heavy metals include cadmium (Cd), lead (Pb), mercury (Hg), arsenic (As), and chromium (Cr) (Fig. 5.1) (Festa and Thiele 2011).

Treatment with chelators is one of the first therapeutic strategies described against Pb poisoning (Flora et al. 2012). Flavonoids (Quercetin, Acid and Alpha-Lipoic), vitamins (B, C, and E), trace elements (Zn, Cu, Se, Co, and so on), and plant-based antioxidants (Garlic, Saffron, and curcumin) can be considered an important protective agent against toxicity, and oxidative stress induced by Pb (Hsu and Guo 2002).

Thanks to its antioxidant and chelating power, curcumin is one of the medicinal plants that has shown a powerful pharmacological and therapeutic power against various pathologies, namely inflammatory, cancerous, viral, fungal, mutagenic, depressive, and neurodegenerative diseases (Mahmoud et al. 2018; Shi et al. 2018).

5.2 Lead Neurotoxicity and Neurodegenerative Diseases

5.2.1 General Information on Lead

Pb is one of the most ubiquitous and widespread heavy metals in the environment, thus the oldest and most widely used by man. It is used in various industrial fields mainly (metallurgy, battery manufacturing, paints, electrical equipment, and so on) (Kapusta and Sobczyk 2015; Lalor et al. 2006; Lanphear et al. 1998), which

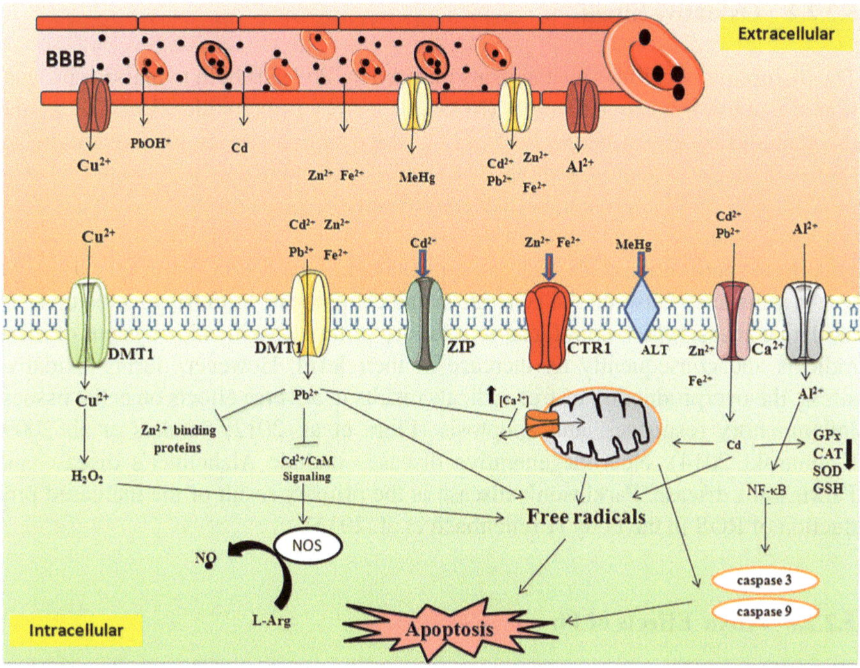

Fig. 5.1 Toxicity of metal elements

increases the range of exposure in humans to this metal. The absorption of Pb can occur through different routes (Digestive, Respiratory, and Dermal) (ATSDR 2007).

5.2.2 *Mechanisms of Pb Toxicity*

5.2.2.1 Ionic Mechanism

Due to its structural and chemical properties similar to those of calcium (Ca2+), zinc (Zn2+), magnesium (Mg2+), and iron (Fe2+), Pb exerts its toxic effects by mimicking their physiological effects and competing with them for their sites of action for which it often has a high affinity, where it binds to specific sites on Ca2 + −binding proteins (Flora et al. 2006, 2012; Mazzolini et al. 2001). Pb, therefore, will cause disruption of various Ca2 + −dependent enzymatic and cellular mechanisms (Bressler et al. 1999). In addition, Pb has the ability to influence mitochondrial function by inhibiting Ca2+ uptake and storage mechanisms, which are also the regulatory elements of calcium homeostasis (Pounds et al. 1991). Consequently, Pb can cause an alteration of intracellular signaling mechanisms, and consequently be the origin of the inhibition of the release of certain neurotransmitters at the presynaptic terminals, such as acetylcholine (Devoto et al. 2001).

5.2.2.2 Oxidative Stress

The disruption of cellular mechanisms induced by Pb leads both to massive production of reactive oxygen derivatives (ROS), such as hydroperoxides (HO2-), singular oxygen, and hydrogen peroxide (H2O2), which in turn causes oxidative damage, such as lipid peroxidation, protein, and DNA damage. On the other hand, Pb toxicity leads to a depletion of antioxidant enzymes such as superoxide dismutase (SOD) in blood, catalase (CAT), glutathione-S-transferase (GST), glutathione (GSH), and glutathione peroxidase (GPx) (Flora et al. 2006; Han et al. 2005; Sanders et al. 2009).

An imbalance between the generation and chelation of ROS triggers the onset of oxidative stress, which could lead to a failure of the detoxification system of these radicals and consequently an increase in their level. However, during oxidative stress, the overproduction of free radicals results in adverse effects on cells, tissues, inflammatory responses, and apoptosis (Flora et al. 2012; Sanders et al. 2009; Szymanski 2014). Neurodegenerative diseases include Alzheimer's disease and Parkinson's disease. Parkinson's disease is the primary result of the increased production of ROS in the body (Breitenbach et al. 2013).

5.2.2.3 Toxic Effects of Pb

Exposure to Pb may cause several symptoms of poisoning in various organs and systems that may be affected: renal, bone, reproductive and endocrine, cardiovascular, hematological, immunological, developmental, genetic, carcinogenic, and neurological (ATSDR 2007).

5.2.3 At the Peripheral Level

5.2.3.1 Digestive Effects

At the level of the digestive tract, Pb intoxication induces digestive disorders characterized by the following symptoms: anorexia, nausea, vomiting, weight loss, acute abdominal pain, constipation, or diarrheal episodes, and bloating (Goyer 1993; Kelada et al. 2001). The exact mechanism of these various induced symptoms is unknown.

5.2.3.2 Renal System

Upon cumulative exposure to Pb, the kidneys remain among the targeted vital organs, which may be affected by acute or chronic nephropathy with several health complications, namely, impairment of the tubular absorption and reabsorption mechanism, degeneration of the tubular epithelium, renal dysfunction, renal failure, hypertension, and hyperuricemia (Flora et al. 2012; Tamegart et al. 2021). In comparison, acute nephropathy caused by Pb intoxication is reversible by the cessation of exposure to the intoxication, whereas chronic nephropathy is irreversible and develops over a long period of time due to prolonged exposure (Lin et al. 2001; Odigie et al. 2004).

5.2.3.3 Bones

After Pb exposure, bone remains a primary storage site in the body (Renner 2010). In adults, 85–95% of absorbed Pb is stored in bones, whereas children store only 70%. Studies have shown that Pb poisoning impairs bone growth while reducing its calcium and phosphorus contents (Hamilton and O'Flaherty 1994).

5.2.3.4 Reproductive and Endocrine System

The reproductive system also can be affected by Pb exposure, contributing to decreased libido, abnormal spermatogenesis, infertility, serum testosterone changes, abnormal prostate function, miscarriage, and early birth, in both men and women. Previous studies have indicated that lead can cause peritubular testicular fibrosis, a reduction in the sperm count (Flora et al. 2012).

Other studies have shown that Pb can induce the disruption of luteinizing hormone (LH) regulation, decreased testosterone synthesis, delayed prenatal follicle maturation and increased ovarian destruction, and reduced primordial follicle numbers in female rodent pups (Apostoli and Catalani 2010; Assi et al. 2016).

5.2.3.5 Cardiovascular System

Pb intoxication is classically accompanied by arterial hypertension but rather also associated with other deleterious cardiovascular effects such as strokes, peripheral arterial disease (arteriosclerosis and atherosclerosis), coronary artery disease that leads to poor cardiovascular function, and abnormalities such as cardiac arrhythmias and left ventricular hypertrophy (Bonde et al. 2002).

5.2.3.6 Hematological Parameters

As soon as Pb reaches the bloodstream, 99% of the metal binds to red blood cells and is subsequently transported and distributed throughout the body, only 1% is found in plasma in the ionized form. The saturation of Pb binding sites in the red blood cells could lead to an increased bioavailability of this metal in plasma (Ambrose et al. 2000). The alteration of heme synthesis and the destruction of red blood cells are significantly linked to Pb toxicity by preventing the action of d-aminolevulinic acid dehydratase (ALAD), ferrochelatase aminolevulinic acid synthetase (ALAS), aminolevulinic acid (ALA), and other mitochondrial enzymes, which reduces the life of circulating erythrocytes, due to persistent instability of the cell membrane and leads to anemia (Bouchard et al. 2009; Flora et al. 2012). This anemia caused by Pb intoxication can be the consequence of either an acute exposure that results in so-called hemolytic anemia, or a chronic intoxication (Vij and Dhundasi 2009).

5.2.3.7 Carcinogenic Effects

Occupational exposure to high doses of Pb in workers and its inorganic compounds could result in the development of bronchial, renal, and gastric tumors. Based on experimental results and data, the International Agency for Research on Cancer has considered Pb as carcinogenic to humans (Higginson and DeVita Jr 1980).

5.2.3.8 In the Central Nervous System

Bradbury and Deane (1993) demonstrated a linear uptake of Pb at the level of different brain structures in adult rats over a four-hour period, confirming the ability of Pb to cross the blood-brain barrier (BBB) (Bradbury and Deane 1993).

5.2.3.9 Effects on the Glial System

Glial cells are the main component of the central nervous system (CNS), which do not have the capacity to generate or transport an action potential, so they have the power to regenerate even in adulthood (Tortora and Grabowski 2001). They have several functions, namely survival and maintenance of neuronal function and differentiation (Hunter-Schaedle 1997) as well as the metabolism of neurotransmitters (Araque et al. 2001). At the level of the CNS, astrocytes constitute the most responsive type, where they surround neurons and ensure their contact with blood vessels, and thanks to their feet they form a kind of blood brain barrier capable of regulating exchanges between the bloodstream and the nervous tissues (Gazzaniga et al. 2000).

During chronic Pb intoxication, astrocytes show reactive gliosis in the frontal cortex (Sansar et al. 2011). These astroglial reactions are the result of the accumulation and sequestration of this metal in order to reduce its bioavailability (Holtzman et al. 1987) and consequently protect the neurons that are more sensitive (Tiffany-Castiglioni 1993). After the saturation of glial cells with this metal, it is released in a continuous manner, contributing to the intoxication of neurons and even neighboring astroglia (Sabbar 2013).

5.2.4 *Effects on the Neuronal System*

5.2.4.1 Monoaminergic Neurotransmission and pb Poisoning

Monoamines are neurotransmitters involved in the formation of neuronal circuits during brain development. Numerous studies have investigated the effects of Pb on the monoaminergic system. Exposure to Pb blocks the Ca2 + −dependent secretory activity of acetylcholine (ACh), dopamine (DA), and monoacid neurotransmitters (Devoto et al. 2001; Lasley et al. 1999). Whereas at low doses Pb stimulates the spontaneous release of neurotransmitters (Bressler et al. 1999; Tamegart et al. 2019a, b).

Cory-Slechta et al. (1998) showed that intoxication by Pb acetate of the order of 50 to 150 ppm potentiates the release of DA at the level of nucleus accumbens, following the injection of potassium chloride (KCl) (Cory-Slechta et al. 1998; Zuch et al. 1998), as well as an increase in the sensitivity of D2 receptors to its agonists while reducing the effects of its antagonists. On the other hand, chronic studies have shown a decrease in tyrosine hydroxylase (TH) levels in the substantia nigra pars compacta (SNc) and motor cortex in adult rats exposed to Pb in drinking water, 5 g/l (Sansar et al. 2011).

Serotonergic (5-HTergic) neurotransmission also, as for the Dopaminergic one, shows alterations during Pb exposure. Rats acutely and chronically exposed to a dose of 3 and 5 g/l respectively, which showed enhanced serotonin (5-HT) immunoreactivity in the dorsal raphe (Benammi et al. 2014; Sansar et al. 2011; Tamegart et al. 2021). In addition, developmental exposure to Pb also alters monoamine oxidase (MAO) activity in the brain of rats and mice (Devi et al. 2005).

5.2.4.2 GABAergic Neurotransmission and Pb Intoxication

GABA (ϒ aminobutyric acid) is the most common inhibitory neurotransmitter in the CNS. In the brain, we find more the GABAergic system, where its role is paramount (Scheel-Krüger et al. 1981). Otto and Murata (1993) showed that Pb exposure decreases GABA transport and affects the properties of ionotropic GABA receptors. While other studies have shown a decrease in cortical GABA levels associated with developmental Pb intoxication (Leret et al. 2002; Otto and Murata 1993). Pb intoxication may be one of the possible causes of alterations in the GABAergic system that is involved in various behaviors (Leret et al. 2002).

5.2.4.3 Glutamatergic Neurotransmission and Pb Poisoning

Glutamate (Glut) is among the essential amino acids, it is the main excitatory neurotransmitter of the CNS, of which more than 40% of the excitatory synapses are Glutergic (Meldrum 2000), in addition, Glut is involved in many metabolic pathways and physiological functions namely; synaptic plasticity, cognitive functions, and development of the nervous system (Kaneko and Mizuno 1994; Yano et al. 1998). Exposure to Pb leads to a decrease in the glutamate levels in the brain, hippocampus, and cerebellum (Leret et al. 2002; Verstraeten et al. 2008), other studies have shown the inhibition of glutamatergic NMDA (N-methyl-D-aspartate) receptors in the hippocampal neurons involved in long-term potentiation and memory in fetuses (Jett et al. 1997), the AMPA (α-amino 3-hydroxy-5-methyl-4- isoxazole propionic acid) receptor, is also affected by Pb, through reduced binding of Glut to its AMPA receptor in the cortex and hippocampus (Chen et al. 2001). Pb-induced alterations in Glutamatergic receptors, namely NMDAs and AMPAs, could explain the adverse effects of this metal on learning and memory (Chen et al. 2001).

5.3 Aluminum Neurotoxicity and Neurodegenerative Diseases

5.3.1 *Toxic Effects of Al*

Aluminium constitutes 8% of the earth's crust. It is the third most abundant element behind oxygen and silicon. It is found only in the combined form, in the oxide form, in bauxite and complex aluminosilicates such as mica and feldspar. In contrast with its richness in the Earth's crust, its concentration in the ocean is less than 1 μg of aluminum per liter, this low level could be caused by the accumulation of aluminum and silica by diatoms (Hydes 1977).

In recent years, the effects of industrialization have disturbed the balance between nature and man. The brutal and massive use of heavy metals, including aluminum, has gradually led to the appearance of new risks, still poorly evaluated, including neurological disorders.

Although the risks of intoxication by low levels of heavy metals are not yet taken very seriously by the majority of conventional medical practitioners, the danger definitely exists. New research has shown that these metals accumulate in the body and, because of this, can have very harmful effects. Aluminum, in particular, increases the risk of cardiovascular and kidney diseases, hypertension and cancer, and decreases fertility and memory (Keitel et al. 2008). The bioavailability of strange metals in the body generally contributes to the appearance of certain species called reactive oxygen species, which include all radical derivatives of oxygen but also other highly reactive non-radical compounds. Some reactive nitrogen species (RNS) are sometimes mentioned as belonging to this classification because they have an oxygen atom and behave in a similar way (generally radical species, which have a high oxidizing power, which are generated and regulated by the organism) to ROS in the face of oxidative stress (Neag et al. 2012) (Fig. 5.1).

The appearance of severe oxidative stress can therefore unbalance the natural antioxidant defense systems, which induces a general disorder that can affect the antioxidant enzyme barriers and even other systems such as the central nervous system, including their function in Alzheimer's disease, Parkinson's disease, and amyotrophic lateral sclerosis (Lou Gehrig's disease) ... (Roos et al. 2006).

Aluminum affects various enzymes, disrupts different metabolic pathways, and has a necrotizing effect. Aluminum stock sites are the first target of its toxic action. Particularly on the central nervous system, it exerts several actions such as a change in the permeability of the BBB, this alteration allows the passage of L-glutamic acid and citrate at the cerebral level following the membrane permeability. It results in the formation of aluminum-L-glutamate complexes, which accentuate the alteration of the BBB as well as that of the erythrocyte membrane (McLachlan et al. 1991) (Fig. 5.1).

Aluminum modifies the activities of choline acetylase and acetylcholinesterase (non-competitive inhibition by a conformational change), and consequently, disrupts the absorption of choline by the neurons.

It also reduces the concentrations of inositol triphosphate and cyclic AMP. This mechanism is directly responsible for influencing calcium-dependent signaling and processes through the inhibition of voltage-gated channels for calcium entry into neurons and its extrusion from the neuronal cytosol by the Mg2 + −dependent ATPase.

It can modulate the expression of messenger RNAs (mRNAs) and protein synthesis by accumulation in the nucleus of neurons, followed by binding to chromatin and alteration of the expression of mRNAs coding for certain pro-inflammatory cytokines involved in neurotoxic mechanisms and neurofibrillary degeneration (Kwon et al. 2013).

Aluminum is highly neurotoxic and, at high concentrations, can inhibit prenatal and postnatal brain development in humans and experimental animals (Yumoto et al. 2009). At a young age, the brain maybe the most sensitive target organ to aluminum.

A number of neurological events have been attributed to aluminum poisoning in humans. These include memory loss, tremors, saccadic movements, coordination disorders, slow motor movements, loss of curiosity, ataxia, myoclonic remorse, and generalized convulsions with epileptic status (Sanderson et al. 1982; Zatta et al. 2003).

5.3.2 Al and Neurodegenerative Diseases

Neuropathological disorders associated with high brain aluminum include senile and dementia of the Alzheimer's type, Down's syndrome with manifested Alzheimer's disease, Guam and Kiipeninsula, amytropic lateral sclerosis (ALS) (spinal cord and brain), multiple sclerosis (MS), Parkinson's dementia with neurofibrillar degeneration, dialysis encephalopathy, striatonigral syndrome, alcoholic dementia with uneven demyelination, senile plaques of Alzheimer's disease, and aging brain (Zatta et al. 2003). The amount of activated astrocytes is high in AD and this phenomenon is particularly pronounced in senile plaques (Zatta et al. 2003). In the hippocampus of AD patients, there is increased expression of pro-inflammatory genes (Colangelo et al. 2002).

AD is associated with the deposition of toxic amyloid β-peptide (Aβ), produced from the amyloid-β precursor protein (AβPP) via proteolytic processes (Fig. 5.1). The brain of Alzheimer's disease contains reactive microglia, producing pro-inflammatory cytokines and acute phase proteins, in the vicinity of Aβ-containing neuritic plaques (Styren et al. 1998). Nasally instilled aluminum nanoparticles have also been shown to reach the brain via the olfactory tract, resulting in the activation of pro- and anti-inflammatory kinases (Kwon et al. 2013).

Aluminum salts induce Aβ aggregation in vitro and treatment of Aβ PP-overexpressing transgenic mice with Al salts in drinking water results in oxidative stress, Aβ deposition, and plaque formation in the brain (Praticò et al. 2002). However, two studies on aluminum and the promotion of Alzheimer's pathology or behavior have challenged this conclusion (Akiyama et al. 2012; Poirier et al. 2011). An emerging generalization may be that the behavioral effects of aluminum are clearer in normal aging animals, while they are more difficult to be detected in mutant animal strains, which are previously predisposed to plaque formation and memory deficits (Ribes et al. 2008).

Epidemiological, neuropathological, and biochemical research suggests a potential link between aluminum neurotoxicity and the pathogenesis of AD. Nevertheless, the link between those pathological disorders and aluminum is still controverted. These similarities provide emerging and important evidence for the role of aluminum in the genesis of Alzheimer's disease (Rondeau et al. 2009) (Fig. 5.1):

- Aluminum may inhibit the metabolism of tetrahydrobiopterin, which is important in the production of several neurotransmitters, and tetrahydrobiopterin metabolism is known to be decreased in Alzheimer's disease.
- A number of studies from different groups have indicated that aluminum is associated with the pathological lesions found at autopsy in the brains of patients with Alzheimer's disease. Aluminum appears to co-localize at the very center of the pathological plaque and in neurons supporting the typical neurofibrillary degeneration. The fact that aluminum is found at the heart of such lesions may suggest a causative role.
- Many studies were reported in cats and other mammals using aluminum as one of the means of behavioral (and neuropathological) model induction for AD.
- In a relatively small number of brains of patients dying of aluminum encephalopathy, neuropathological changes have been identified that are similar although not identical to those in Alzheimer's disease. The difference in confounding may result from the fact that acute aluminum intoxication in hemodialysis patients is very significant compared to the chronic or supposed lifetime intoxication of patients destined to develop Alzheimer's syndrome from aluminum poisoning, in

the context of renal and other failures, causal clinical effects that are often similar to those seen in Alzheimer's disease.

5.4 Mn and Copper Neurotoxicity and Neurodegenerative Diseases

Manganese (Mn) is an important mineral that helps cells operate normally (Erikson et al. 2005). This oligo-element is thought to be a cofactor of a wide range of biological processes involved in cellular homeostasis and survival, as well as during organ and tissue development, such as bone formation, blood sugar regulation, reproduction, lipid, protein, and carbohydrate metabolism, immune response, and antioxidant defense. Mn is implicated in the activity of multiple enzymes, including pyruvate decarboxylase, glutamine synthetase (GS), serine/threonine protein phosphatase I, arginase, and Mn-SOD, which are essential for neurotransmitter production and metabolism, as well as glial and neuronal function (Erikson and Aschner 2003; Horning et al. 2015).

Mn is primarily obtained through the environment and food supply. The environmental pathways include air, soil, and water, while the daily intake compromises rice, seafood, fruits, and vegetables. For adults, including pregnant women, the European Food Safety Authority (EFSA) recommended a daily intake of 3.0 mg. Mn intake, whether excessive or insufficient, has an impact on human health. It is assumed to enter cells by passive diffusion or active transport through the divalent metal transporter 1 (DMT-1) (Aschner et al. 2007). Manganese deficiency, though infrequent, can disrupt physiological function and produce a variety of health problems (Aschner et al. 2002); however, this is not clearly understood.

High levels of Mn exposures were first linked to negative health effects in working contexts. Oral exposure to Mn-contaminated drinking water caused similar signs such as tremors, difficulties walking, and facial muscle spasms, which appeared later in life. Human exposure to Mn is clearly linked to neurological consequences, according to a growing body of data (Williams et al. 2012). Mn easily crosses the blood-brain barrier in a growing fetus or newborn. Neurological effects are divided into three types: mental, behavioral, and motor functions (Wahlberg et al. 2018).

The disruption of mitochondrial activity, which depletes adenosine triphosphate (ATP) and stimulates the creation of reactive oxygen species (ROS), has been proposed to cause damage to the central nervous system, contributing to neurodegeneration through a vicious cycle in the case of homeostasis alteration of this element. Metals promote apoptotic and/or necrotic cell death through these pathways (Garza-Lombó et al. 2018).

Mn comes in a variety of chemical forms, such as oxidation states (Mn^{2+}, Mn^{3+}, Mn^{4+}, Mn^{6+}, and Mn^{7+}), salts (sulfate and gluconate), and chelates (aspartate, fumarate, and succinate). The two most prevalent forms found in the human body are divalent (Mn^{2+}) and trivalent (Mn^{3+}) forms. Mn^{2+} is the most common type of Mn found in metalloenzymes (Horning et al. 2015). Mn^{2+} is efficiently converted to Mn^{3+} by ceruloplasmin. Mn^{2+} has been revealed to enter the central nervous system via a transferrin receptor (TfR)-mediated pathway. Mn^{2+}, unlike Mn^{3+}, is easily transported into the brain as a free ion or as a non-specific protein-bound species (Aschner and Dorman 2006).

Manganese is easily concentrated in the brain, particularly in the basal ganglia (Kabata et al. 1989), and can result in Manganism, an irreversible neurological illness that resembles Parkinson's disease. Manganism is a Parkinson's-like extrapyramidal movement condition that produces devastating motor and cognitive abnormalities. It is caused by a neurodegenerative process (Chen et al. 2015)that induces dystonia, a neurological indication associated with damage to the globus pallidus. The occurrence of global bradykinesia and extensive rigidity is a similarity between manganism and Parkinson's disease (Calne et al. 1994). The cellular and molecular mechanism behind Mn's neurotoxicity has yet to be fully elucidated.

5.4.1 Mechanisms of Mn Neurotoxicity

5.4.1.1 Alzheimer Type II Astrocytosis

Astrocytes are specialized glial cells found in the central nervous system that are necessary for optimum neuronal activation. Astrocytes store 50 times more Mn than neurons, making them an important target cell for Mn transport and the most vulnerable neural cell to Mn toxicity (Zwingmann et al. 2003). Since the major homeostatic regulator for Mn in the brain are astrocytes, they are considered an important target cell for Mn transport and storage. Alzheimer type II astrocytosis refers to the pathological changes in astrocytes caused by Mn poisoning.

According to results from experimental research of primary microglia and astrocytes, microglia directly collect Mn and generate a mixed inflammatory phenotype characterized by the release of cytokines and CC chemokines. The findings reveal that products from Mn-activated microglia are required for Mn-exposed astrocytes neuroinflammation, and that NF-κB-dependent TNF- α release from microglia is a crucial signaling event in microglia that regulates these glial-glial interactions (Kirkley et al. 2017; Tjalkens et al. 2017).

Furthermore, as astrocytes are responsible for the elimination of synaptic glutamate (Glu), Mn accumulation in them has been demonstrated to disrupt Glu homeostasis and induce excitatory neurotoxicity (Butterworth et al. 1995). Glial cells are a key target of Mn in the brain, both for the sequestration of the metal and for activating inflammatory signaling pathways that damage neurons by overproducing a variety of inflammatory cytokines, as well as reactive oxygen and nitrogen species (ROS and RNS) (Tjalkens et al. 2017).

5.4.1.2 Mitochondrial Dysfunction and Oxidative Stress

Mitochondria is the principal storage site for intracellular Mn, and increased Mn levels in this organelle can interfere directly with cellular ATP production (Martinez-Finley et al. 2013). Once within the mitochondria, Mn can block the enzymatic complexes of the mitochondrial electron transport chain, especially complexes I and II. Interestingly, mitochondrial complex II is a significant mediator of Mn-induced hydrogen peroxide synthesis and, as a result, excessive ROS that leads to oxidative stress. Mn has also been shown to affect Ca2+ homeostasis in the mitochondria by blocking efflux and hence increasing its levels in this organelle (Martinez-Finley et al. 2013). As a result of the oxidative stress caused by the elevated Mn levels, the mitochondrial permeability transition (MPT) is induced. The opening of these (MPT) pores is responsible for mediating apoptosis, chromatin condensation, and DNA fragmentation (Milatovic et al. 2011).

Excessive Mn uptake in astrocytes and consequent Mn storage in the mitochondria may affect energy metabolism, increase oxidative damage, and alter astrocytic–neuronal communication, resulting in changes in excitatory and inhibitory regulation and contributing to secondary Glu-mediated toxicity (Filipov and Dodd 2012).

5.4.1.3 Mn-Induced Alterations in Neurotransmitter Systems

Excessive Mn accumulation can mimic Parkinson's disease symptoms. Both Mn2+ and Mn3+ catalyze the auto-oxidation of dopamine (DA) and produce reactive species, contributing to oxidative damage. Chronic Mn exposure is linked to DAergic neurodegeneration in the nigrostriatal area (Benedetto et al. 2009). A study showed, in pretreated mice, the ability of Mn to disrupt dopamine neurotransmission by modifying tyrosine hydroxylase activity in the nigrostriatal pathway, confirming Mn′s potential neurotoxic effect as a risk factor for Parkinson's disease onset (El Fari et al. 2019). The effect of Mn on striatal dopamine (DA) levels, however, is still debated (Benedetto et al. 2009).

Glutamate (Glu) is the most abundant excitatory neurotransmitter and Mn accumulation causes symptoms that are compatible with an excitotoxic mechanism. Excitotoxicity is caused by an excess of Glu in the extracellular space, which can trigger a series of degenerative events downstream (Takeda 2003). Excess Glu is eliminated through cellular absorption, which is mediated by glutamate 1 transporter (GLT1) and glutamate–aspartate transporter (GLAST), both of which are abundant in astrocytes (Tuschl et al. 2013). Mn exposure can reduce astrocytes' ability to eliminate synaptic Glu, increasing its excitotoxicity. Furthermore, changes in energy supply have a significant impact on Glu distribution. As a result, mitochondrial failure must be recognized as a crucial factor in Mn-induced Glu dyshomeostasis mediation. Furthermore, Mn-induced reactive oxygen species production may directly block Glu absorption, resulting in an increase in extracellular Glu levels (Erikson and Aschner 2003; Takeda 2003).

GABA is the most common inhibitory neurotransmitter. Mn promotes GABA release by boosting the GABA content in the caudate nucleus. GABAergic neurons in the striatum are more sensitive to Mn than mesencephalic DAergic neurons in cell culture models. Mn is largely stored in the globus pallidus, however, its effects on GABAergic neurotransmission are debatable (Kwakye et al. 2015). Because Mn intoxication patients frequently exhibit symptoms that are similar to those of Parkinson's disease, the basal ganglia have been investigated as a possible site for Mn toxicity research (Lao et al. 2017). The fundamental effect of Mn on GABAergic systems has yet to be discovered (Erikson and Aschner 2003).

The cholinergic system, which includes acetylcholine (Ach)-secreting neurons and accompanying glia, is involved in cognition, emotions, and/or locomotion. Mn activity is thought to affect a number of cholinergic synaptic functions, including presynaptic choline uptake, quantal release of Ach into the synaptic cleft, post-synaptic binding of Ach to receptors, and acetylcholinesterase-mediated synaptic breakdown. Astrocytic choline transport and astrocytic Ach-binding proteins are also affected by Mn. Thus, symptoms linked with manganism's early psychotic phase, also known as 'manganese madness,' may be linked to a dysfunction of the septohippocampal cholinergic system, which is implicated in both physiological and behavioral stress responses (Peres et al. 2016).

5.4.1.4 Neuroinflammation

Although Mn can directly affect neurons through oxidative damage and mitochondrial dysfunction, manganese has also been shown to increase neurotoxicity through glial cell activation and subsequent release of inflammatory cytokines and non-neuronal ROS. Manganese has been demonstrated to promote neurotoxicity through glial cell activation and subsequent production of inflammatory cytokines and non-neuronal ROS (Filipov and Dodd 2012).

Mn has the ability to cause microglial cells to release interleukin (IL)-1b, IL-6, and tumor necrosis factor (Filipov and Dodd 2012). The presence of astrocytes is required for Mn-induced neuronal damage, indicating the importance of astrocytes in Mn neurotoxicity. Manganese increases ERK-dependent NF-kB signaling via potentiating cytokine-induced NO synthase expression and NO production in astrocytes by activating soluble guanylate cyclase (Tjalkens et al. 2008, 2017).

Overall, the inflammation of glial cells within the basal ganglia, followed by neurotoxic lesion, appears to be an important mechanism of Mn toxicity.

5.5 Conclusion

Metallic trace elements are known for their toxic effects on all aquatic species and mammals. The toxicity of these metals is able to affect the brain and cause neurotoxicity, which has an impact on neurodegenerative diseases as elucidated above.

Each metal has its own mechanism for crossing the blood-brain barrier; some form complexes with proteins and others are transported by specific transporter receptors to the CNS. The choroid plexus also plays an important role in the passage of metal to the CNS, from where it reaches different parts of the brain and induces neurotoxicity through distinct mechanisms. It is also possible that plant extracts containing many phytochemicals form complexes with these metals thereby protecting enzymes or proteins or both.

References

Akiyama H, Hosokawa M, Kametani F, Kondo H, Chiba M, Fukushima M, Tabira T (2012) Long-term oralintake of aluminium or zinc does not accelerate Alzheimer pathology in AβPP and AβPP/tau transgenic mice. Neuropathology 32:390–397

Ambrose TM, Al-Lozi M, Scott MG (2000) Bone lead concentrations assessed by in vivo X-ray fluorescence. Clin Chem 46(8):1171–1178

Apostoli P, Catalani S (2010) Metal ions affecting reproduction and development. Met Ions Toxicol Eff Interact Interdepend Met Ions Life Sci 8:263–303

Araque A, Carmignoto G, Haydon PG (2001) Dynamic signaling between astrocytes and neurons. Annu Rev Physiol 63:795–813

Aschner M, Dorman DC (2006) Manganese : pharmacokinetics and molecular mechanisms of brain uptake. Toxicol Rev 25(3):147–154. https://doi.org/10.2165/00139709-200625030-00002

Aschner M, Guilarte TR, Schneider JS, Zheng W (2007) Manganese : recent advances in understanding its transport and neurotoxicity. Toxicol Appl Pharmacol 221(2):131–147. https://doi.org/10.1016/j.taap.2007.03.001

Aschner M, Shanker G, Erikson K, Yang J, Mutkus LA (2002) The uptake of manganese in brain endothelial cultures. Neurotoxicology 23(2):165–168. https://doi.org/10.1016/s0161-813x(02)00056-6

Assi MA, Hezmee MNM, Haron AW, Sabri MYM, Rajion MA (2016) The detrimental effects of lead on human and animal health. Vet world 9:660

ATSDR, U.S., 2007. Toxicological profile for lead. US Dep. Heal. Hum. Serv. 1, 582

Benammi H, El Hiba O, Romane A, Gamrani H (2014) A blunted anxiolytic like effect of curcumin against acute lead induced anxiety in rat: involvement of serotonin. Acta Histochem 116:920–925

Benedetto A, Au C, Aschner M (2009) Manganese-induced dopaminergic neurodegeneration : insights into mechanisms and genetics shared with Parkinson's disease. Chem Rev 109(10):4862–4884. https://doi.org/10.1021/cr800536y

Bonde JP, Joffe M, Apostoli P, Dale A, Kiss P, Spano M, Caruso F, Giwercman A, Bisanti L, Porru S (2002) Sperm count and chromatin structure in men exposed to inorganic lead: lowest adverse effect levels. Occup Environ Med 59:234–242

Bouchard MF, Bellinger DC, Weuve J, Matthews-Bellinger J, Gilman SE, Wright RO, Schwartz J, Weisskopf MG (2009) Blood lead levels and major depressive disorder, panic disorder, and generalized anxiety disorder in US young adults. Arch Gen Psychiatry 66:1313–1319

Bradbury MW, Deane R (1993) Permeability of the blood-brain barrier to lead. Neurotoxicology 14:131–136

Breitenbach M, Ralser M, Perrone GG, Iglseder B, Rinnerthaler M, Dawes IW (2013) Oxidative stress and neurodegeneration: the yeast model system. Front Biosci (Landmark Ed) 18:1174–1193

Bressler J, Kim K, Chakraborti T, Goldstein G (1999) Molecular mechanisms of lead neurotoxicity. Neurochem Res 24:595–600

Butterworth RF, Spahr L, Fontaine S, Layrargues GP (1995) Manganese toxicity, dopaminergic dysfunction and hepatic encephalopathy. Metab Brain Dis 10(4):259–267. https://doi.org/10.1007/BF02109357

Calne DB, Chu NS, Huang CC, Lu CS, Olanow W (1994) Manganism and idiopathic parkinsonism : similarities and differences. Neurology 44(9):1583–1586. https://doi.org/10.1212/wnl.44.9.1583

Chen H-H, Ma T, Ho K (2001) Effects of developmental lead exposure on inhibitory avoidance learning and glutamate receptors in rats. Environ Toxicol Pharmacol 9:185–191

Chen P, Chakraborty S, Mukhopadhyay S, Lee E, Paoliello MMB, Bowman AB, Aschner M (2015) Manganese homeostasis in the nervous system. J Neurochem 134(4):601–610. https://doi.org/10.1111/jnc.13170

Cohen JA, Trojano M, Mowry EM, Uitdehaag BMJ, Reingold SC, Marrie RA (2020) Leveraging real-world data to investigate multiple sclerosis disease behavior, prognosis, and treatment. Mult Scler J 26:23–37

Colangelo V, Schurr J, Ball MJ, Pelaez RP, Bazan NG, Lukiw WJ (2002) Gene expression profiling of 12633 genes in Alzheimer hippocampal CA1: transcription and neurotrophic factor down-regulation and up-regulation of apoptotic and pro-inflammatory signaling. J Neurosci Res 70:462–473

Cory-Slechta DA, O'Mara DJ, Brockel BJ (1998) Nucleus accumbens dopaminergic medication of fixed interval schedule-controlled behavior and its modulation by low-level lead exposure. J Pharmacol Exp Ther 286:794–805

Devi CB, Reddy GH, Prasanthi RPJ, Chetty CS, Reddy GR (2005) Developmental lead exposure alters mitochondrial monoamine oxidase and synaptosomal catecholamine levels in rat brain. Int J Dev Neurosci 23:375–381

Devoto P, Flore G, Ibba A, Fratta W, Pani L (2001) Lead intoxication during intrauterine life and lactation but not during adulthood reduces nucleus accumbens dopamine release as studied by brain microdialysis. Toxicol Lett 121:199–206

El Fari R, Abbaoui A, Bourziq A, Zroudi M, Draoui A, El Khiat A, Belkouch M, Elgot A, Gamrani H (2019) Neuroprotective effects of docosahexaenoic acid against sub-acute manganese intoxication induced dopaminergic and motor disorders in mice. J Chem Neuroanat 102:101686. https://doi.org/10.1016/j.jchemneu.2019.101686

Erikson KM, Aschner M (2003) Manganese neurotoxicity and glutamate-GABA interaction. Neurochem Int 43(4-5):475–480. https://doi.org/10.1016/s0197-0186(03)00037-8

Erikson KM, Syversen T, Aschner JL, Aschner M (2005) Interactions between excessive manganese exposures and dietary iron-deficiency in neurodegeneration. Environ Toxicol Pharmacol 19(3):415–421. https://doi.org/10.1016/j.etap.2004.12.053

Festa RA, Thiele DJ (2011) Copper: an essential metal in biology. Curr Biol 21:R877–R883

Filipov NM, Dodd CA (2012) Role of glial cells in manganese neurotoxicity. J Appl Toxicol 32(5):310–317. https://doi.org/10.1002/jat.1762

Flora G, Gupta D, Tiwari A (2012) Toxicity of lead: a review with recent updates. Interdiscip Toxicol 5:47–58

Flora SJS, Flora G, Saxena G (2006) Environmental occurrence, health effects and management of lead poisoning, in: Lead. Elsevier, pp. 158–228

Ganguly U, Chakrabarti SS, Kaur U, Mukherjee A, Chakrabarti S (2018) Alpha-synuclein, proteotoxicity and Parkinson's disease: search for neuroprotective therapy. Curr Neuropharmacol 16:1086–1097

Garza-Lombó C, Posadas Y, Quintanar L, & Gonsebatt ME (2018) Neurotoxicity linked to dysfunctional metal ion homeostasis and xenobiotic metal exposure : redox signaling and oxidative stress—PubMed. https://pubmed.ncbi.nlm.nih.gov/29402131/

Gazzaniga MS, Ivry RB, Mangun GR (2000) Neurosciences cognitives: la biologie de l'esprit. De Boeck Supérieur

Goyer RA (1993) Lead toxicity: current concerns. Environ Health Perspect 100:177–187

Hamilton JD, O'Flaherty EJ (1994) Effects of lead exposure on skeletal development in rats. Fundam Appl Toxicol 22:594–604

Han SG, Kim Y, Kashon ML, Pack DL, Castranova V, Vallyathan V (2005) Correlates of oxidative stress and free-radical activity in serum from asymptomatic shipyard welders. Am J Respir Crit Care Med 172:1541–1548

Higginson J, DeVita VT Jr (1980) IARC monographs on the evaluation of carcinogenic risk of chemicals to humans. Am Ind Hyg Assoc J 41:A26–A28

Holtzman D, Olson JE, DeVries C, Bensch K (1987) Lead toxicity in primary cultured cerebral astrocytes and cerebellar granular neurons. Toxicol Appl Pharmacol 89:211–225

Horning KJ, Caito SW, Tipps KG, Bowman AB, Aschner M (2015) Manganese is essential for neuronal health. Annu Rev Nutr 35:71–108. https://doi.org/10.1146/annurev-nutr-071714-034419

Hsu P-C, Guo YL (2002) Antioxidant nutrients and lead toxicity. Toxicology 180:33–44

Hunter-Schaedle KE (1997) Radial glial cell development and transformation are disturbed in reeler forebrain. J Neurobiol 33:459–472

Hydes DJ (1977) Reaction of aluminum at clay surfaces, in: Transactions-American Geophysical Union. Amer Geophysical Union 2000 Florida Ave NW, WASHINGTON, DC 20009, p. 1165

Jett DA, Kuhlmann AC, Guilarte TR (1997) Intrahippocampal administration of lead (Pb) impairs performance of rats in the Morris water maze. Pharmacol Biochem Behav 57:263–269

Kabata H, Matsuda A, Yokoi K, Kimura M, & Itokawa Y (1989) [The effect of the dosage and route of manganese administration on manganese concentration in rat brain] Nihon Eiseigaku Zasshi. Japanese J Hyg, 44(2), 667–672. https://doi.org/10.1265/jjh.44.667

Kaneko T, Mizuno N (1994) Glutamate-synthesizing enzymes in GABAergic neurons of the neocortex: a double immunofluorescence study in the rat. Neuroscience 61:839–849

Kapusta P, Sobczyk Ł (2015) Effects of heavy metal pollution from mining and smelting on enchytraeid communities under different land management and soil conditions. Sci Total Environ 536:517–526

Keitel WA, Campbell JD, Treanor JJ, Walter EB, Patel SM, He F, Noah DL, Hill H (2008) Safety and immunogenicity of an inactivated influenza A/H5N1 vaccine given with or without aluminum hydroxide to healthy adults: results of a phase I–II randomized clinical trial. J Infect Dis 198:1309–1316

Kelada SN, Shelton E, Kaufmann RB, Khoury MJ (2001) δ-Aminolevulinic acid dehydratase genotype and lead toxicity: a HuGE review. Am J Epidemiol 154:1–13

Kirkley KS, Popichak KA, Afzali MF, Legare ME, Tjalkens RB (2017) Microglia amplify inflammatory activation of astrocytes in manganese neurotoxicity. J Neuroinflammation 14(1):99. https://doi.org/10.1186/s12974-017-0871-0

Kwakye GF, Paoliello MMB, Mukhopadhyay S, Bowman AB, Aschner M (2015) Manganese-induced Parkinsonism and Parkinson's disease : shared and distinguishable features. Int J Environ Res Public Health 12(7):7519–7540. https://doi.org/10.3390/ijerph120707519

Kwon JT, Seo GB, Jo E, Kim HM, Shim I, Lee BW, Yoon BI, Kim P, Choi K (2013) Aluminum nanoparticles induce ERK and p38MAPK activation in rat brain. Toxicol Res 29:181–185

Lalor GC, Vutchkov MK, Bryan ST, Christie CDC, Donaldson D, Young J, Chambers S (2006) Acute lead poisoning associated with backyard lead smelting in Jamaica. West Indian Med J 55:394–398

Lanphear BP, Matte TD, Rogers J, Clickner RP, Dietz B, Bornschein RL, Succop P, Mahaffey KR, Dixon S, Galke W (1998) The contribution of lead-contaminated house dust and residential soil to children's blood lead levels: a pooled analysis of 12 epidemiologic studies. Environ Res 79:51–68

Lao Y, Dion L-A, Gilbert G, Bouchard MF, Rocha G, Wang Y, Leporé N, Saint-Amour D (2017) Mapping the basal ganglia alterations in children chronically exposed to manganese. Sci Rep 7:41804. https://doi.org/10.1038/srep41804

Lasley SM, Green MC, Gilbert ME (1999) Influence of exposure period on in vivo hippocampal glutamate and GABA release in rats chronically exposed to lead. Neurotoxicology 20:619–629

Leret ML, Garcia-Uceda F, Antonio MT (2002) Effects of maternal lead administration on monoaminergic, GABAergic and glutamatergic systems. Brain Res Bull 58:469–473
Lin J-L, Tan D-T, Hsu K-H, Yu C-C (2001) Environmental lead exposure and progressive renal insufficiency. Arch Intern Med 161:264–271
Mahmoud AA, Abdelrahman A, el Aziz HOA (2018) Protective effect of curcumin on the liver of high fat diet-fed rats. Gene Rep 11:18–22
Martinez-Finley EJ, Gavin CE, Aschner M, Gunter TE (2013) Manganese neurotoxicity and the role of reactive oxygen species. Free Radic Biol Med 62:65–75. https://doi.org/10.1016/j.freeradbiomed.2013.01.032
Mazzolini M, Traverso S, Marchetti C (2001) Multiple pathways of Pb2+ permeation in rat cerebellar granule neurones. J Neurochem 79:407–416
McLachlan DR, Kruck TP, Lukiw WJ, Krishnan SS (1991) Would decreased aluminum ingestion reduce the incidence of Alzheimer's disease? C Can Med Assoc J 145:793
Meldrum BS (2000) Glutamate as a neurotransmitter in the brain: review of physiology and pathology. J Nutr 130:1007S–1015S
Milatovic D, Gupta RC, Yu Y, Zaja-Milatovic S, Aschner M (2011) Protective effects of antioxidants and anti-inflammatory agents against manganese-induced oxidative damage and neuronal injury. Toxicol Appl Pharmacol 256(3):219–226. https://doi.org/10.1016/j.taap.2011.06.001
Neag A, Favier V, Bigot R, Pop M (2012) Microstructure and flow behaviour during backward extrusion of semi-solid 7075 aluminium alloy. J Mater Process Technol 212:1472–1480
Odigie IP, Ladipo CO, Ettarh RR, Izegbu MC (2004) Effect of chronic exposure to low levels of lead on renal function and renal ultrastructure in SD rats. Niger J Physiol Sci 19:27–32
Otto D, Murata K (1993) Summary of workshop III: evoked potentials. Environ Res 60:79–81
Peres TV, Schettinger MRC, Chen P, Carvalho F, Avila DS, Bowman AB, Aschner M (2016) Manganese-induced neurotoxicity : a review of its behavioral consequences and neuroprotective strategies. BMC Pharmacol Toxicol 17(1):57. https://doi.org/10.1186/s40360-016-0099-0
Poirier J, Semple H, Davies J, Lapointe R, Dziwenka M, Hiltz M, Mujibi D (2011) Double-blind, vehicle-controlled randomized twelve-month neurodevelopmental toxicity study of common aluminum salts in the rat. Neuroscience 193:338–362
Pounds JG, Long GJ, Rosen JF (1991) Cellular and molecular toxicity of lead in bone. Environ Health Perspect 91:17–32
Praticò D, Uryu K, Sung S, Tang S, Trojanowski JQ, Lee VMY (2002) Aluminum modulates brain amyloidosis through oxidative stress in APP transgenic mice. FASEB J 16(9):1138–1140
Renner R (2010) Exposure on tap: drinking water as an overlooked source of lead.
Ribes D, Colomina MT, Vicens P, Domingo JL (2008) Effects of oral aluminum exposure on behavior and neurogenesis in a transgenic mouse model of Alzheimer's disease. Exp Neurol 214:293–300
Rondeau V, Jacqmin-Gadda H, Commenges D, Helmer C, Dartigues J-F (2009) Aluminum and silica in drinking water and the risk of Alzheimer's disease or cognitive decline: findings from 15-year follow-up of the PAQUID cohort. Am J Epidemiol 169:489–496
Roos PM, Vesterberg O, Nordberg M (2006) Metals in motor neuron diseases. Exp Biol Med 231:1481–1487
Sabbar M (2013) Conséquences de la toxicité du plomb sur l'activité des ganglions de la base et les rythmes circadiens chez le rat: Approches électrophysiologique, neurochimique et anatomo-fonctionnelle
Sanders T, Liu Y, Buchner V, Tchounwou PB (2009) Neurotoxic effects and biomarkers of lead exposure: a review. Rev Environ Health 24:15–46
Sanderson CL, McLachlan DRC, De Boni U (1982) Altered steroid induced puffing by chromatin bound aluminum in a polytene chromosome of the blackfly Simulium Vittatum. Can J Genet Cytol 24:27–36
Sansar W, Ahboucha S, Gamrani H (2011) Chronic lead intoxication affects glial and neural systems and induces hypoactivity in adult rat. Acta Histochem 113:601–607

Scheel-Krüger J, Magelund G, Olianas MC (1981) Role of GABA in the striatal output system: globus pallidus, nucleus entopeduncularis, substantia nigra and nucleus subthalamicus. Adv Biochem Psychopharmacol 30:165

Shi L-Y, Zhang L, Li H, Liu T-L, Lai J-C, Wu Z-B, Qin J (2018) Protective effects of curcumin on acrolein-induced neurotoxicity in HT22 mouse hippocampal cells. Pharmacol Rep 70:1040–1046

Styren SD, Kamboh MI, Dekosky ST (1998) Expression of differential immune factors in temporal cortex and cerebellum: the role of α-1-antichymotrypsin, apolipoprotein E, and reactive glia in the progression of Alzheimer's disease. J Comp Neurol 396(4):511–520

Szymanski M (2014) Molecular mechanisms of lead toxicity. Biotechnol J Biotechnol Comput Biol Bionanotechnol:95

Takeda A (2003) Manganese action in brain function. Brain Res Brain Res Rev 41(1):79–87. https://doi.org/10.1016/s0165-0173(02)00234-5

Tamegart L, Abbaoui A, Bouyatas MM, Gamrani H (2021) Lead (Pb) exposure induces physiological alterations in the serotoninergic and vasopressin systems causing anxiogenic-like behavior in Meriones shawi: assessment of BDMC as a neuroprotective compound for Pb-neurotoxicity and kidney damages. J Trace Elem Med Biol 65:126722

Tamegart L, Abbaoui A, El Khiat A, Bouyatas MM, Gamrani H (2019a) Altered nigrostriatal dopaminergic and noradrenergic system prompted by systemic lead toxicity versus a treatment by curcumin-III in the desert rodent Meriones shawi. C R Biol 342:192–198

Tamegart L, Abbaoui A, Makbal R, Zroudi M, Bouizgarne B, Bouyatas MM, Gamrani H (2019b) Crocus sativus restores dopaminergic and noradrenergic damages induced by lead in Meriones shawi: a possible link with Parkinson's disease. Acta Histochem 121:171–181

Tiffany-Castiglioni E (1993) Cell culture models for lead toxicity in neuronal and glial cells. Neurotoxicology 14:513–536

Tjalkens RB, Liu X, Mohl B, Wright T, Moreno JA, Carbone DL, Safe S (2008) The peroxisome proliferator-activated receptor-gamma agonist 1,1-bis(3′-indolyl)-1-(p-trifluoromethylphenyl) methane suppresses manganese-induced production of nitric oxide in astrocytes and inhibits apoptosis in cocultured PC12 cells. J Neurosci Res 86(3):618–629. https://doi.org/10.1002/jnr.21524

Tjalkens RB, Popichak KA, Kirkley KA (2017) Inflammatory activation of microglia and astrocytes in manganese neurotoxicity. Adv Neurobiol 18:159–181. https://doi.org/10.1007/978-3-319-60189-2_8

Tortora GJ, Grabowski SR (2001) Principes d'anatomie et de physiologie. De Boeck Supérieur

Tuschl K, Mills PB, Clayton PT (2013) Manganese and the brain. Int Rev Neurobiol 110:277–312. https://doi.org/10.1016/B978-0-12-410502-7.00013-2

Verstraeten SV, Aimo L, Oteiza PI (2008) Aluminium and lead: molecular mechanisms of brain toxicity. Arch Toxicol 82:789–802

Vij AG, Dhundasi SA (2009) Hemopoietic, hemostatic and mutagenic effects of lead and possible prevention by zinc and vitamin C. Al Ameen J Med Sci 2:27–36

Wahlberg K, Arora M, Curtin A, Curtin P, Wright RO, Smith DR, Lucchini RG, Broberg K, Austin C (2018) Polymorphisms in manganese transporters show developmental stage and sex specific associations with manganese concentrations in primary teeth. Neurotoxicology 64:103–109. https://doi.org/10.1016/j.neuro.2017.09.003

Williams M, Todd GD, Roney N, Crawford J, Coles C, McClure PR, Garey JD, Zaccaria K, & Citra M (2012) Toxicological profile for manganese. Agency for Toxic Substances and Disease Registry (US). http://www.ncbi.nlm.nih.gov/books/NBK158872/

Willis AW, Evanoff BA, Lian M, Galarza A, Wegrzyn A, Schootman M, Racette BA (2010) Metal emissions and urban incident Parkinson disease: a community health study of Medicare beneficiaries by using geographic information systems. Am J Epidemiol 172:1357–1363

Yano S, Tokumitsu H, Soderling TR (1998) Calcium promotes cell survival through CaM-K kinase activation of the protein-kinase-B pathway. Nature 396:584

Yumoto S, Kakimi S, Ohsaki A, Ishikawa A (2009) Demonstration of aluminum in amyloid fibers in the cores of senile plaques in the brains of patients with Alzheimer's disease. J Inorg Biochem 103:1579–1584

Zatta P, Lucchini R, van Rensburg SJ, Taylor A (2003) The role of metals in neurodegenerative processes: aluminum, manganese, and zinc. Brain Res Bull 62:15–28

Zerrouki K (2017) L'effet antioxydant de quelques plantes médicinales sur la neurotoxicité et les maladies neurodégénérative dues aux métaux lourds (Etude expérimentale chez la souris)

Zuch CL, O'Mara DJ, Cory-Slechta DA (1998) Low-level lead exposure selectively enhances dopamine overflow in nucleus accumbens: AnIn VivoElectrochemistry time course assessment. Toxicol Appl Pharmacol 150:174–185

Zwingmann C, Leibfritz D, Hazell AS (2003) Energy metabolism in astrocytes and neurons treated with manganese : relation among cell-specific energy failure, glucose metabolism, and intercellular trafficking using multinuclear NMR-spectroscopic analysis. J Cereb Blood Flow Metab 23(6):756–771. https://doi.org/10.1097/01.WCB.0000056062.25434.4D

Chapter 6
Edible Bird's Nest: Seeing the Unseen

Wael Mohamed

Abstract Cognitive enhancement is the growth or augmentation of the mind's core capacities via the improvement or expansion of internal or external information processing systems. Edible bird nest (EBN) is a natural dietary ingredient formed from the saliva of edible velvet nests. Supplementation with EBN has been shown to boost brain growth in animals. During times of fast brain development, notably in premature newborns, the bioactivity and nutritional value of EBN are substantial. Nevertheless, the effect of EBN on learning and memory regulation is uncertain. This chapter attempts to show the neuroprotective properties of EBN and its possible cognitive-enhancing effects.

Keywords Cognitive enhancement · Edible bird nest · Nutrition · Neuroprotection

6.1 Introductory Considerations

A photographic memory, the capacity to retain and recall enormous quantities of knowledge instantaneously, and similar dreams have been the stuff of science fiction for a long time. There is a reasonable explanation for this. Our grasp of neuronal intelligence structures and fundamental engrams is still in its infancy. Memory formation is believed to include both the synthesis of proteins in brain cells and the reorganization of the three-dimensional connections among dendritic processes. In the case of long-term potentiation, for instance, we have established the electrophysiological basis for conditioned cellular behavior as well as some of the associated cellular signaling mechanisms. Despite this, and despite the vast quantity of knowledge we currently possess, we are no closer to achieving our long-held

W. Mohamed (✉)
Department of Basic Medical Sciences, Kulliyyah of Medicine, International Islamic University Malaysia (IIUM), Kuantan, Malaysia

Clinical Pharmacology Department, Menoufia Medical School, Menoufia University, Menoufia, Egypt

W. Mohamed, R. Sandhir (eds.), *Trace Elements in Brain Health and Diseases*, Nutritional Neurosciences, https://doi.org/10.1007/978-981-99-1513-2_6

objective of increased cognitive ability. However, drugs and natural items promise to achieve the same results as media advertisements. While some of these dietary brain boosters are usually seen as safe, the notion of greatly enhancing cognitive ability, memory, or learning in cognitively normal adults through conventional pharmacological treatments is sometimes considered as unethical. Alzheimer's disease has been the subject of the most rigorous scientific research, with an emphasis on the reversal of dementia-related cognitive impairment. Drugs that have been found to be beneficial in animal models of dementia or in people with the condition are often also effective in healthy persons. As a result, medicines and their natural product equivalents will very probably be used to enhance cognitive performance in a variety of circumstances. The purpose of this review was to highlight the available cognitive enhancers (CEs) and emphasize the positive effects of Edible Bird Nest (EBN) on the human brain. Whether you believe it or not, natural goods have a large and growing fan base among the general public. The West is witnessing the most rapid development, but even in less economically developed nations, demand remains solid. This argument is bolstered by the almost universal, but false belief that a product must be safe if it is natural. EBN is one of the naturally occurring compounds believed to have a beneficial influence on cognitive performance, and this review will analyze it.

6.2 Cognitive Enhancers

6.2.1 Introduction and Context

Cognitive enhancement is the growth or augmentation of the mind's core capacities via the improvement or expansion of internal or external information processing systems (Sandberg and Bostrom 2006). This category includes brain treatments that improve executive functions such as thinking and decision-making by enhancing attention, focus, and information processing (Glannon 2015). Sometimes CEs are referred to as memory-enhancing drugs or nootropics.

According to Glannon (2015), the three most important principles in cognitive improvement are increase, reduce, and optimize. The first of these principles is growth. The first concept considers brain treatments to be enhancements when they improve a function through enhancing the brain's capacity to conduct its usual functions. According to the second principle, some functions may be enhanced by reducing their operational scope and the effect of their activities. It is a plan for enhancing the overall welfare. Under some conditions, a person's diminished ability or function may contribute to their overall well-being: more is not always better and sometimes less is more (Earp et al. 2014). For instance, the medication methylphenidate (Ritalin) may help certain individuals do a particular cognitive task better because it reduces the content of their thoughts, enabling them to ignore distracting stimuli and focus more successfully on the task at hand. According to the third concept, improved information processing functions are defined as any intervention aimed at optimizing a certain class of information processing functions; cognitive processes, which are physically performed by the human brain (Metzinger and Hildt 2011).

It is conceivable to categorize as therapeutic an intervention designed to treat a particular illness or a cognitive subsystem impairment. A modification is a form of intervention that enhances a subsystem in a manner other than by repairing a defect or correcting a particular fault. In practice, it is sometimes difficult to differentiate between treatment and augmentation, and the difference may have little practical value to begin with. When a person's natural memory is damaged, cognitive enhancement may result in a memory that is poorer than that of a person with an identifiable illness, such as early-stage Alzheimer's disease, but who retains a fair degree of memory function. A cognitively enhanced person does not necessarily possess very high (or even superhuman) cognitive abilities on their own. Cognitively enhanced persons are those who have benefitted from an intervention that enhances the performance of a cognitive subsystem without treating a particular ailment or condition.

The range of cognitive improvements includes medical therapies, psychological interventions (such as taught "tricks" or mental strategies), and upgrades to external institutional and technical systems that facilitate cognition (Table 6.1). CEs are distinguished from other forms of treatment by the fact that they increase basic cognitive capacities as opposed to specialized, tightly defined skills or domain-specific knowledge.

Table 6.1 Cognitive enhancement strategies

Strategies	Examples of interventions
Biochemical	• Nutrition • Natural remedies • Recreational drugs • Pharmaceuticals • Body derivatives
Physical	• Implants • Electrical stimulation • Magnetic stimulation • Optical stimulation • Acoustic stimulation • Gadgets
Behavioral	• Computer training • Mnemonics • 2nd language learning • Meditation • Sleep • Physical exercise

Adapted from Dresler et al. (2019)

6.2.2 Mechanism of Action of CE

Numerous individuals feel that a certain neurotransmitter system correlates to a particular cognitive function. This is not always the case. Therefore, it is conventional to believe that cognitive enhancement would operate in the same manner. Dopamine is associated with working memory and attention (Robbins and Arnsten 2009), while serotonin is associated with emotional functions (Cools et al. 2008). In contrast, serotonin, noradrenaline, and acetylcholine may have an effect on working memory (Luciana et al. 2001). Dopamine is implicated in reinforcement learning in response to rewards (Schultz 2002), whereas serotonin seems to influence reinforcement learning in response to aversive stimuli (Dayan and Huys 2009).

In addition, it is known that neurotransmitters work via a number of receptor systems; hence, they may exhibit a range of behaviors. For instance, the actions of dopamine on D1 receptors may be diametrically opposite to those on D2 receptors (Floresco and Magyar 2006). There are 17 different receptor systems for the neurotransmitter serotonin. Furthermore, depending on where it is delivered, dopamine may have quite distinct effects in various brain areas and even within different portions of the human basal ganglia (Clatworthy et al. 2009). Other neurotransmitters, such as glutamate in the nucleus accumbens, which is situated inside the hippocampus, may also exert localized regulation (Wise 2004). It is thus plausible that interactions across neuromodulatory systems serve as a means of controlling some of their effects. Based on these considerations, simple conceptualizations linking a certain neurotransmitter to a single cognitive function are unlikely to be effective.

According to the most recent study, there is growing evidence that various neurotransmitters may have distinct modes of action when they are delivered in a tonic, sustained fashion as compared to a phasic one (Aston-Jones and Cohen 2005; Sarter et al. 2009). For instance, the first activation of noradrenergic cells in the locus coeruleus changes depending on the individual's level of wakefulness or arousal. It seems that phasic activation of these cells is associated with appropriate responses to environmental stimuli, but only when tonic activity is low (Aston-Jones and Cohen 2005). As a consequence of phasic activation, changes in the global concentrations of a neurotransmitter may alter the capacity to react to external stimuli mediated by the same neurotransmitter.

What is the method through which the existing enhancer medications acquire their beneficial effects? Mementophenidate, dextroamphetamine, and modafinil are examples of medicines that may assist persons with normal cognitive abilities increase their attention, focus, and other working memory-related skills. Decision-making requires the capacity to store and retrieve information in a timely manner, which includes the ability to save and retrieve information fast. In scientific studies, it has been established that methylphenidate, by increasing dopamine levels in the brain, may result in a modest improvement in these behaviors for the majority of individuals. Short-term studies have shown that individuals with considerably lower baseline working memory are more likely to benefit from this therapy than those with much greater baseline working memory (de Jongh et al. 2008; Farah et al.

2004). Several cognitive functions have deteriorated in some of the later people. Children with attention deficit hyperactivity disorder (ADHD), who use methylphenidate and other stimulant medicines, tend to do better in school than children with the same disease who do not take medication (Scheffler et al. 2009). According to the study, there is no proof that these treatments have a significant effect on the academic performance of children who do not suffer from these conditions.

Methylphenidate seems to create an inverted dose–response curve, according to experiments with healthy participants. Moderate dosages enhance mental performance, however, higher dosages may or may not have an impact on mental performance (de Jongh et al. 2008). Some evidence suggests that methylphenidate increases executive functions in novel tasks, but has a detrimental impact on executive functions in learned tasks (de Jongh et al. 2008). According to the results of a recent experiment in which researchers used transcranial electrical stimulation to test learning and application of mathematical information, they discovered that stimulating a region of the subject's prefrontal cortex hindered their ability to learn new information but enhanced their ability to apply previously acquired knowledge. By activating a portion of the parietal cortex, researchers were able to enhance learning while lowering the capacity to apply newly acquired knowledge (Iuculano and Cohen Kadosh 2013). These studies illustrate that the increase of some mental processes by medicine or electrical stimulation may come at the price of the enhancement of other mental functions. Cognitive trade-offs may need to be weighed while choosing whether or not to improve. As a consequence of the research, it was determined that the concept of augmentative development had flaws. According to their results, there are limitations to how much cognitive function may be enhanced and there are optimal levels of cognitive performance.

Amphetamines promote the release of catecholamines from presynaptic nerve terminals, which raises the concentrations of dopamine and norepinephrine in the synaptic space. They employ competitive inhibition to restrict the reuptake of norepinephrine and dopamine into the presynaptic cell. The release of norepinephrine affects both alpha and beta adrenergic receptor sites (Denzer et al. 2019).

Modafinil inhibits the hypothalamus' capacity to induce sleep by activating dopamine, which raises norepinephrine and histamine levels. Even when sleep-deprived, users of the medication report being more alert and attentive. Despite the fact that it is prescribed for sleep problems, airline pilots with normal sleep–wake cycles have taken it to remain awake on lengthy transcontinental flights. It may also be advantageous to students preparing for exams or writing papers, as well as to researchers writing grant proposals, since it permits them to forgo sleep and devote more time to these activities while awake. Young, healthy volunteers were given 100–200 mg of modafinil, and their levels of attention, spatial planning, and visual pattern recognition were significantly greater than those of the controls (Müller et al. 2004). Before taking action, the capacity to perform a comprehensive analysis of a situation facilitates decision-making. Similar to prior studies on methylphenidate, the effects of modafinil on working memory were more prominent among participants with inferior cognitive capacities at the time of the assessment in this study. Specifically, it enhanced accuracy on a test of sustained attention for adults with low intellect (but still above-average intelligence) and a poor working memory baseline. Those with a higher IQ and stronger working memory at baseline were either unaffected by modafinil or performed worse than those who did not take the drug. Similar to other psychostimulants, modafinil's effects are confined to certain cognitive processes, vary from individual to individual, and are not always advantageous.

It has been shown that Propranolol (Inderal) has an indirect influence on cognition. This beta-adrenergic receptor antagonist, often known as a beta-blocker, reduces the reaction of the circulatory system to adrenaline and noradrenaline. Propranolol may prevent or lessen the symptoms of a scared or anxious reaction to stimuli by lowering the sensitivity of receptors in the body and brain to the chemicals that generate these sensations. Those who are anxious about playing music or making a public speech may utilize this strategy to avoid being distracted by, among other stress-related symptoms, a racing heart or heavy perspiration. Consequently, the medicine may aid you in focusing on the current work and finishing it more swiftly.

In certain instances, using medication to increase concentrate on a given task may have detrimental implications. A study of the cognitive responses of older and younger individuals to a battery of cognitive tests revealed that having a wide attention span and concentrating less on a given activity over time may facilitate the transfer of knowledge from one context to another (Carson et al. 2003). The majority of individuals will gain from being able to think more creatively and solve issues effectively as a consequence. Long-term use of methylphenidate and other drugs with a similar mechanism of action increases certain cognitive abilities, but reduces the capacity to modify behavior and adapt to new conditions. Whether or not there are trade-offs connected with long-term usage may depend on the dose and whether the drug was taken on a regular basis or as required. Contrary to the beliefs of some experts, the findings of a study on mathematical knowledge utilizing transcranial electrical stimulation indicate that for many individuals, consistent use of psychostimulants or non-invasive brain stimulation may assist them enhance certain talents while hampering others. This is, however, an open empirical topic that will not be answered until a substantial number of controlled studies on the cognitive effects of medicines or procedures have been completed.

When it comes to enhancing one's cognitive skills, memory has received considerable attention. While there are many more forms of memory, it is fascinating to consider the possibility of expanding the ability to store and recall more episodic memories of events and semantic memories of facts, ideas, and general environmental information. Memory encoding and consolidation cannot occur without protein production in the hippocampus and neighboring areas of the medial temporal lobes of the brain. According to the notion, drugs that enhance protein synthesis and long-term potentiation via strengthening memory synapses have the potential to boost our capacity to remember more information and experiences. There are several medications that may achieve this. One such family of drugs is the ampakines, which interact with glutaminergic AMPA and NMDA receptors, which play a crucial role in learning. Some persons believe that enhancing episodic and semantic memory storage and retrieval can increase our knowledge and help us to utilize our working memory more effectively; nevertheless, memory augmentation has limits that will be explained below. After a certain point, it may become difficult to recall and retain information from previous experiences. According to the author, adding to them will hinder rather than help one's capacity to build and implement action plans (Hauskeller 2014).

It is challenging to address questions concerning the specificity of cognitive regulation in clinical populations. Parkinson's disease dementia (PDD) and dementia with Lewy bodies are now often treated with acetylcholinesterase inhibitors (AChEIs), which include, among others, rivastigmine and donepezil; (DLB). Numerous clinical studies have shown that these medicines have relatively minor positive effects on cognitive screening tests administered at the bedside (Arvanitakis et al. 2019; Kabir et al. 2020). In addition to other cognitive capacities, sensitive computerized cognitive tests have demonstrated broad enhancements in the areas of attention, working memory, and episodic memory (Tahami Monfared et al. 2020; Wesnes et al. 2005). Although it is conceivable that these positive effects of AChEIs are mediated by a single mechanism, such as increased arousal, it is doubtful that this is the case (Edgar et al. 2009; Naismith et al. 2011). Even the modulatory effects of AChEIs in healthy patients create the same difficulty, which is why it is the topic of our study (Repantis et al. 2010). Donepezil, for instance, has been shown to enhance episodic memory in healthy young volunteers (Chuah et al. 2009), as well as verbal memory in healthy older individuals (FitzGerald et al. 2008). Many of these medications, when provided to Alzheimer's patients, suppress AChE activity in the central nervous system, therefore preventing the disruption of cholinergic neurotransmission. NMDA receptor antagonism and maybe a decrease in the formation of beta-amyloid plaques in the brain are among the reasons by which other drugs enhance cognitive function in Alzheimer's disease patients.

6.2.3 Indications and Applications of CE

In the past decade, a number of pharmacological therapies for increasing cognitive function in a broad spectrum of brain illnesses have been explored and licensed for use in clinical practice (Kokiko and Hamm 2007). Methylphenidate and atomoxetine, which regulate the noradrenergic and dopaminergic systems, are gaining popularity in the treatment of developmental disorders such as attention deficit hyperactivity disorder (ADHD) and hyperactivity (Caye et al. 2018; Posner et al. 2020). AChEIs and memantine [an N-methyl-D-aspartate (NMDA) receptor antagonist] have become routine therapies for neurodegenerative disorders such as Alzheimer's and Parkinson's disease in recent years (Arvanitakis et al. 2019; Olivares et al. 2012; Szeto and Lewis 2016).

Although cognitive impairments are distinct from positive (e.g., hallucinations and delusions) and negative (e.g., lack of emotion and speech) symptoms in chronic mental diseases such as schizophrenia, available antipsychotic drugs have little or no impact on cognitive deficits. Consequently, a variety of medications are being studied for their potential to enhance cognitive performance in this disease (Harvey and Bowie 2012). It is also being investigated if cognitive impairments caused by a stroke might be alleviated (Barfejani et al. 2020; Szelenberger et al. 2020). Many CEs target the neuromodulatory systems of the brain, including the cholinergic, dopaminergic, noradrenergic, and serotonergic systems, which rise from the brain-stem nuclei to innervate both cortical and subcortical systems.

They have had a significant impact, sparking interest in cognitive improvement not only among patients with brain disorders but also among healthy individuals, despite the fact that the majority of the claimed positive effects of such medications are small in magnitude and highly variable between individuals (Table 6.2). Students who want to improve their grades, military personnel who need to stay awake for extended missions, older adults who are concerned about cognitive decline, and even university academics who want to maintain their performance, all use substances like methylphenidate and modafinil to achieve their objectives (Greely et al. 2008; Husain and Mehta 2011; Wilms et al. 2019).

6.2.4 Available CEs and Adverse Effects

The sector of medical enhancement is anticipated to have the greatest number of safety concerns. Due to the fact that the current medical risk system compares treatment risk to the expected benefit of reduced morbidity risk from successful treatment, it is extremely risk-averse with regard to enhancements that do not reduce morbidity risk and whose utility to the patient is entirely non-therapeutic, highly subjective, and context-dependent, as well as enhancements that do not reduce morbidity risk and are entirely non-therapeutic, highly subjective, and context-dependent. There are, however, demonstrable instances of a separate risk paradigm, such as the usage of cosmetic surgery.

Even if the therapy does not prevent or minimize morbidity, the widespread view is that patient autonomy trumps any modest medical risks. Similarly, in the case of medicinal CEs, the user may be able to judge if the advantages exceed the risks based on both medical advice and her own evaluations of how the intervention would affect her own objectives and way of life. It is likely that continuous use of a

Table 6.2 Medication and supplement recommendations for CEs

Cognitive enhancers	Recommended usage
Methylphenidate, amphetamine	Attention deficit hyperactivity disorder, wake-promoting agent
Modafinil	Wake-promoting agent
Atomoxetine, reboxetine	Attention deficit hyperactivity disorder, depression
Donepezil, galantamine, rivastigmine (AChEI)	Alzheimer's disease, Parkinson disease dementia (PDD), Dementia with Lewy bodies (DLB)
Memantine	Alzheimer's disease

pharmaceutical that enhances cognition may lead to medical issues in addition to its intended effects. As a side effect of its intended impact, using a memory booster, for instance, may result in an increase in the amount of little "junk" memories being retained. This might be undesirable (Bostrom and Sandberg 2009). In part because it is sometimes hard to forecast in advance all of the potential hazards connected with long-term usage, users may expect limited advice from medical professionals. Medical professionals are not always able to determine if the advantages exceed the hazards for a particular patient.

As with any other treatment, the use of medications to enhance cognition might have adverse consequences on organs and systems other than the brain. For instance, both AChEIs and methylphenidate usually induce stomach discomfort or nausea, forcing patients to quit usage (Veroniki et al. 2016). These negative consequences have the potential to wipe out any positive benefits of the drug on overall performance, and they should be considered by anybody contemplating the non-medical use of such medications. From the standpoint of cognitive neuroscience, it is more relevant that some medications may impair certain areas of cognition while concurrently improving others in the same person.

Rivastigmine may increase learning on a physical task and the capacity to create connections between symbols and numbers in healthy older adults, but it can impair verbal and visual episodic memory in people with dementia or Alzheimer's disease (Wezenberg et al. 2005). In a similar manner, it has been shown that the dopamine agonist bromocriptine improves spatial working memory while decreasing probabilistic reversal learning in young participants (Mehta et al. 2001). This finding is consistent with research on Parkinson's disease patients: dopaminergic medication improves working memory and task-set switching ability while impairing reverse learning and memory (Cools et al. 2001; Swainson et al. 2000). It is hypothesized that dopamine overload in ventral striatal regions involved in the latter and dopamine replenishment in dorsal striatal areas needed for the former are required to create these paradoxical outcomes (Dagher and Robbins 2009; Swainson et al. 2000). In Parkinson's disease, dopaminergic drug dosages adequate to restore motor function and certain aspects of cognition may also exacerbate other symptoms of the condition.

It has been shown that certain Parkinson's disease patients using dopaminergic agonists, such as benzodiazepines, acquire impulsive behaviors including gambling, excessive shopping, and sexual promiscuity (Weintraub et al. 2006, 2010). As previously mentioned, in Parkinson's disease, such behavior is frequently associated with dyskinesias, which are involuntary movements caused by excessive dopaminergic activation (Voon et al. 2009), supporting the hypothesis that such impulse control issues may be linked to a "overdose" of particular basal ganglia regions. Important to note is that decreasing the dosage of dopaminergic medications may often lessen impulsivity. These results suggest that dopamine agonists used to treat Parkinson's disease may have a broad range of cognitive and behavioral consequences, both positive and negative.

6.2.5 Acceptability to Society and Moral Considerations

Traditional values, perceived naturalness, and perceived directness of their method of action all impact how well-cognitive improvement therapies are received by the general public, and these distinctions are essentially independent of the precise cognitive skill advantages they give. Numerous presently problematic enhancement therapies, such as brain stimulation and medications, have a substantially longer acceptability history than enhancement interventions with a very short acceptance history, such as meditation and diet (Bergström and Lynöe 2008). It is considered that "natural" remedies, such as sleep and exercise, are more effective than technical breakthroughs (Caviola and Faber 2015). In addition, whether the mode of action is perceived as psychologically mediated or more biologically direct, affecting the brain indirectly through the senses or more directly through the cranium or metabolism, plays a role in social acceptance: if an enhancement intervention, such as intensive cognitive or physical training, requires extended efforts or is perceived as a quick and easy shortcut to the same goal, as in the case of smart pills or brain stimulants, social acceptance will be lower. Despite the fact that views based on purely intuitive factors such as tradition, naturalness, or directness frequently rely on cognitive biases rather than rational argument (Caviola et al. 2014), a negative social perception for whatever reason may generate indirect psychological costs for users, which may influence rational evaluations of the respective enhancement intervention (Faulmuller et al. 2013).

A research examines if the use of CEs in academic settings is acceptable in the eyes of professionals (pharmacists, physicians, nurses, lawyers, and accountants) and whether they feel it is appropriate (Ram et al. 2020). Even if the majority of individuals feel that permitting students to take CEs for cognitive enhancement is unfair, there is still uncertainty over the safety and efficacy of CEs when recommended by a physician. Moreover, professionals' attitudes toward CEs varied, which may be attributable, among other things, to differences in self-identity fundamental values and extrinsic aspects of acceptability within the profession in balancing the elements of opportunity, justice, and authenticity in cognitive enhancement (Ram et al. 2020).

Consequently, one of the most controversial questions in the ethical debate is whether enhancement therapies vary merely in terms of their results, such as their advantages and side effects, or if they differ in terms of their mechanism of action (Racine et al. 2021; Bublitz and Merkel 2012). Many individuals feel that an ethical difference should be made between upgrades that are active in the sense that they involve people's participation and those that are more passive (Pavarini et al. 2018). Diverse perspectives on cognitive improvement are prominent in various (sub)cultures, with a more favorable attitude toward cognitive enhancement therapies in Asia (Macer 2012) or among younger individuals, for instance, being more prevalent in Asia (Singhammer 2012). In empirical examinations of perspectives on cognitive enhancement treatments, medical safety, coercion, and justice were shown to be the most significant issues, with non-users expressing more worries about medical safety and fairness than users (Schelle et al. 2014). In light of the fact that drug prohibition is founded not only on the potential toxicity of chemicals, but also on historical circumstances, the difference between soft and hard enhancers seems rather tenuous (Liakoni et al. 2015).

The objective of the intervention is another factor that influences the social acceptability of cognitive enhancement therapy. In its broadest use, the word "cognitive enhancement" refers to any activity or intervention that improves cognitive capacity, regardless of the explicit goal of such development. Increasing the intensity of traits deemed less fundamental to self-identity (Riis et al. 2008) and increasing the intensity of traits in individuals with cognitive impairments or low-performance baselines appears to be more tolerant than increasing the intensity of traits in normal or high achievers (Medaglia et al. 2019; Sabini and Monterosso 2005). In this instance, it is conceivable to establish at least four distinct objectives, each of which leads to a distinct research strategy and ethical evaluation of present or prospective augmentation alternatives (Allen and Strand 2015).

The idea of cognitive improvement that focuses on normal medical or psychological treatments for pathological problems is one of the least troublesome. Cognitive enhancement therapies that attempt to prevent or reduce cognitive decline associated with healthy aging are inextricably intertwined (Pieramico et al. 2014). Cognitive enhancement strategies that attempt to boost cognition in fully healthy persons while remaining within the normal cognitive boundaries tend to be less frequently accepted. The most popular and morally most controversial aspect of cognitive enhancement is the endeavor to raise cognitive capabilities above and beyond normal function, as typified by the cliche of high-functioning students or managers using smart pills to augment their overall performance.

Aside from the distinctions between improving impaired versus healthy cognition, another distinction in the goals of cognitive enhancement touches on the ultimate deed of an enhancement intervention: given the central role that cognitive capacities play in defining humans as a species, it is tempting to view the improvement of these capacities as a valuable end in and of itself. In contrast, the majority of philosophical and theological approaches emphasize ideals that are only tangentially connected to cognitive performance, such as living an overall better or more meaningful life, rather than objective cognitive performance indicators. Theoretically, human augmentation may be achieved at the population level via sociopolitical changes, as opposed to just through individual cognitive or neurological processes as is the case now (Lucke and Partridge 2012; Schleim 2014).

More than a decade has passed since the ethical issues raised by cognitive enhancement were first discussed (Farah et al. 2004), and numerous experts have raised numerous ethical concerns, including potential risks to mental and physical health (Zaami et al. 2020). Despite evidence that CEs may benefit persons with a range of neuropsychiatric disorders, including Alzheimer's disease and schizophrenia (Grossberg et al. 2019; Lees et al. 2017), their use by healthy people poses ethical, clinical, and legal problems. In addition, a realistic policy framework is required. The use of CEs in healthy people may have multiple benefits, including: reducing social disparity by mitigating negative environmental effects (such as poverty) on the brain; improving the performance of people who must perform at the highest levels in all situations (such as surgeons or pilots); and reducing the use of CEs by self-medicating individuals with undiagnosed disorders (such as ADHD).

This is excellent news, but it is also concerning since the long-term safety and effectiveness of these medications in healthy individuals are yet unclear. While some CEs have been studied and research data on their mechanism of action and potential benefit is available, the majority of CEs have yet to be fully defined and comprehended in terms of their mechanism of action, beneficial effects, and potential adverse effects, according to the most recent research. Moreover, it seems that the effects of CEs (if any) are transient, lasting only until they are metabolized and eliminated from the body (de Jongh et al. 2008). Due to the immaturity of their brains, children and teenagers are especially vulnerable to the adverse effects of many of these drugs, which may lead to addiction and a variety of other adverse effects. It is also essential to include studies that did not provide findings as well as those that demonstrated task-specific deficits (Hall and Lucke 2010).

In order to make informed judgments, the appropriate legislative and regulatory agencies should carefully evaluate the limited evidence of efficacy and any possible negative implications. In 2015, the United States Presidential Commission for the Study of Bioethical Issues (Bioethics Commission) published a report on CE that commented on the most current findings and gave suggestions to doctors. The Australian Alcohol and Drug Foundation has cast doubt on the vast majority of CEs, asserting that scientific research has demonstrated little to no benefit in terms of cognitive improvement in healthy individuals, while the associated side-effects pose significant risks to the public's health and safety (Alcohol and Drug Foundation. Nootropics 2020). There is a need for additional research, but identifying the CEs that are most frequently discussed on the internet will increase clinician awareness of the phenomenon and may aid them in making clinical decisions for patients who are experiencing psychiatric symptoms or physical health problems due to the use of these substances (Barceló et al. 2017; Marchei et al. 2016). Researchers in the area of analytical toxicology may find NPSfinder® useful for focusing their efforts on finding the most recently used medications.

6.3 Edible Bird Nest (EBN)

6.3.1 An Overview and Historical Context

This little insectivorous bird, known as a swiftlet, uses saliva from its specialized salivary glands to construct its nest. Swiftlet species are mostly found in Southeast Asia (Macron 2005; Chua et al. 2016). During the nesting and mating season, male swiftlets produce a valuable natural meal termed EBN from salivary secretion of sticky glycoprotein via sublingual glands. More than 24 species of swiftlets construct nests for their young around the globe. A few are edible, but the vast majority are not. There are three primary swiftlet species that construct EBN: Collocalia, Aerodramus, and Hydrochoun (Wong 2013). *Aerodramus fuciphagus* and *Aerodramus maximus* both build white and black nests, respectively, and are the most exploited and well-known swiftlet species in Malaysia. Completing the mix are feathers and nest-feeding insects, which account for the remaining 70–90% or so *Aerodramus fuciphagus* (Wong 2013; Chua et al. 2016).

Since the Tang dynasty in 681 AD, the Chinese people have regarded EBN as a valuable food and medicine known as the "Caviar of the East" (Macron 2005). In ancient times, only the monarch and the wealthy had access to the EBN soup, which was prepared by double-boiling rock sugar. Despite its reputation as an expensive traditional medicine, it has been used by the Chinese for more than a millennium for its nutritional value and health advantages. As long as people stick to conventional thinking, EBN will continue to be seen as a nutritious food and beauty enhancer that may cure a variety of respiratory and digestive disorders. In addition, it increases the look of aging skin and strengthens the immune system. EBN has been demonstrated to be beneficial for asthma, cough, and stomach ulcers (Kong et al. 1987; Ma and Liu 2012; Hobbs 2004; Macron 2005). Recent research has shown that EBN has antiviral and neuroprotective qualities by inhibiting influenza infection (Haghani et al. 2016, 2017; Xie et al. 2018). EBN has antioxidant, anti-inflammatory, and bone-building effects (Matsukawa et al. 2011; Yida et al. 2014). Due to its therapeutic and palatable properties, EBN is gaining popularity around the globe (Ma and Liu 2012). EBN, a key component in health-enhancing foods, drinks, and cosmetics, has been synthesized utilizing contemporary technology (Kong et al. 1987).

6.3.2 *Active EBN Compounds*

EBN has a unique chemical makeup (Macron 2005: Quek et al. 2018). Protein and carbohydrates are two of the most physiologically active components of EBN, and they are essential for determining the efficiency of the medicine. On average, around 60% of EBN's weight is composed of raw protein. Amino acids, the building blocks of proteins, are essential for the development and regeneration of body cells as well as the synthesis of neurotransmitters, antibodies, and immunoglobulin (Wang 1921; Macron 2005; Chua et al. 2014). The variable composition of EBN amino acids is mostly due to the diverse collecting locations and cave or man-made dwelling types used by EBN (Seow et al. 2016). Researchers discovered that EBN is rich in serine, threonine, and aspartic acids, glutamic acids, prolines, and valines (Babji et al. 2018; Kathan and Weeks 1969). The aromatic amino acid tyrosine, which has antidepressant and analgesic effects, is a key component of white EBN (Macron 2005). This protein, which has the same properties as ovotransferrin from chicken eggs, was identified in EBN. EBN is responsible for children's allergic reactions when they consume it (Macron 2005). The EBN's unique protein and carbohydrate composition is very beneficial to human health (Wang 1921; Xin et al. 2014; Babji et al. 2018; Kathan and Weeks 1969). Research determined that EBN's signature peptide is a mucin-like protein, which might be utilized to categorize EBN based on its color and collection sites (Wong et al. 2017). The protein content of EBN increases as a consequence of its digestion in the stomach and the action of its acidic enzymes (Wong et al. 2018). Carbohydrates are the second most essential component of EBN. 9% of EBN is composed of N-acetylneuraminic acid (sialic acid), Galactosamine, N-acetylglucosamine, and N-acetyl galactosamine, according to Kathan and Weeks, who discovered that 28% of EBN is composed of carbohydrates. Glyconutrients are the compounds responsible for EBN's unique utility (Wong et al. 2017, 2018; Ma and Liu 2012; Chua et al. 2015; Kathan and Weeks 1969).

Sialic acid, which has pharmacological effects on human health, is the single indication that permits the classification of various EBN types (Quek et al. 2018). EBN contains sialic acid in the form of Neu5Ac (Neu5Ac or NANA) (Pozsgay et al. 1987; Guo et al. 2006; Yida et al. 2015a, b; Zhao et al. 2016). It is essential for neuronal proliferation, synaptic transmission, and brain development. A diet rich in sialic acid facilitates both the activation of brain cells and the improvement of cognitive ability (Wang 2009, 2012; Khalida et al. 2019). EBN has a greater concentration than foods rich in sialic acid, such as human milk and chicken egg yolk (Quek et al. 2018). Due to its high concentration, EBN is beneficial for brain growth, influenza prevention, immunological enhancement, cell proliferation, and neurological enhancement (Aswir and Wan Nazaimoon 2011; Hou et al. 2017; Wong et al. 2018; Xie et al. 2018).

6.3.3 EBN's Antioxidant Effects

After hydrolysis, EBN has been identified as an antioxidant with the amino acids contained in EBN. The antioxidant impact of EBN will be improved upon gastrointestinal digestion (Yew et al. 2014; Yida et al. 2014). According to studies conducted by Yida et al., EBN may lessen the risk of hypercoagulation linked with cardiovascular disease (CVD). By lowering oxidative stress, the EBN-treated group was able to enhance the lipid profile, decrease the blood sugar level, and increase total cholesterol in comparison to the control group (Yida ct al. 2015a, b). In 2015, the same research team established the impact of EBN on the oxidative stress caused by a high-fat diet (HFD) in a rat model. Through transcriptional regulation of the expression of hepatic antioxidant genes associated with inflammation, EBN was shown to diminish the oxidative stress and inflammation caused by HFD (Yida et al. 2015a, b). The findings demonstrate that EBN is helpful in preventing inflammation and oxidative stress caused by obesity. In addition, a research by Ghassem et al. indicated that protein hydrolysate of EBN has antioxidant capabilities and may scavenge free radicals (Ghassem et al. 2017). Superoxide dismutase (SOD), estrogen, malondialdehyde, and lipid profile were improved in ovariectomized rats supplemented with EBN for 12 weeks, according to a research with comparable findings (Hou et al. 2015). These results demonstrate the efficacy of EBN in preventing the cardiometabolic disorders brought on by estrogen deprivation. In a Drosophila melanogaster model, Hu et al. examined the anti-aging impact of EBN. The research demonstrated that EBN might lower death rates and lipid peroxidation by boosting the activity of antioxidant enzymes (Hu et al. 2016). Similarly, studies conducted by Albishtue and colleagues to determine the impact of EBN supplementation on uterine function and embryo implantation rate shown that EBN increases anti-oxidant activity and reduces oxidative stress level, which increases embryo implantation rate (Albishtue et al. 2019).

6.3.4 EBN's Effects on Cognition

Nutrition is necessary for a newborn's normal development and growth. Any nutritional deficiency has a significant impact on brain development. It has been shown that the sialic acid-containing glycoprotein EBN enhances brain function. Studies indicate that sialic acid may boost a child's intellect and brain function by increasing the synaptic pathway and ganglioside distribution (Wang 2009, 2012). When sialic acid is consumed as a dietary supplement, a number of genes involved in cognitive development are upregulated in the physiological system (Wang et al. 2007). According to studies done by Xie et al., when EBN was supplied to pregnant and nursing mothers, BDNF and sialic acid levels in the hippocampus rose (Xie et al. 2018). When EBN is injected, the neuronal cell density in the hippocampus CA1, CA2, and CA3 regions increases. By enhancing superoxide dismutase (SOD) and choline acetyltransferase (ChAT) activities while decreasing AChE activity, EBN improved the learning and memory capacity of the offspring (Xie et al. 2018). The BDNF gene attribution in pregnant and breastfeeding female mice revealed that EBN supplementation increased the infants' learning and memory abilities (Mahaq et al. 2020). BDNF expression in the region of the hippocampus may stimulate neurogenesis by boosting mitochondrial biogenesis and neuronal plasticity (Xie et al. 2018; Mahaq et al. 2020). Sialic acid in EBN supplementation increases brain gene expression related to enhanced cognitive performance in the Y maze in both animal generations (Mahaq et al. 2020). Since the quantity of sialic acid in various EBN sources varies, it is currently unclear whether or not EBN supplementation impacts brain gene expression. Mice given varying doses of sialic acid produced from EBN showed improvement in cognitive impairment. When EBN was introduced to the culture of pheochromocytoma and neuroblastoma cells, their proliferation increased (Khalida et al. 2019). Researchers discovered a correlation between brain development and function and the amount of sialic acid in the blood (Khalida et al. 2019). In a prior research, it was shown that EBN improved memory and learning in Wistar rats with neuroinflammation produced by LPS. Memory may be enhanced by inhibiting neuroinflammation using sialic acid's anti-inflammatory properties in EBN, according to research (Careena et al. 2018). In the treatment of cognitive impairment caused by menopause, the excessive use of EBN might be advantageous. According to Zhiping et al., EBN as a natural supplement may relieve the cognitive impairment associated with menopause. EBN decreased estrogen deficiency and the downregulation of genes associated to neurodegeneration in the hippocampus and frontal cortex, according to the findings of this study. Advanced glycation end products (AGEs) associated with estrogen deficiency were significantly decreased by EBN. Additionally, EBN increases antioxidant enzyme activity to decrease oxidative stress in the hippocampus and frontal cortex (Zhiping et al. 2015). The results are consistent with those of a 2017 research which demonstrated that providing EBN to ovariectomized rats increased their cognitive capacities in the hippocampus. By enhancing EBN's activity in the SIRT1 gene in the brain, neuronal plasticity in the hippocampus, which is associated with cognitive ability, might be enhanced. In addition, EBN is a safer alternative to estrogen as a therapeutic agent. Despite the fact that estrogen treatment may boost cognitive capacities, the kidney and liver of the ovariectomized rat may be negatively impacted by it (Hou et al. 2017). EBN glycoprotein has the potential to prevent oxidative damage, which is one of its anti-cancer effects. Multiple investigations have identified the neuroprotective and antioxidant glycoproteins Lactoferrin (LF) and ovotransferrin (OVF) (Ibrahim et al. 2007; Rousseau et al. 2013). When EBN was examined on SHSY5Y cells, Hou and colleagues showed that its antioxidant and protective properties were also owing to its components LF and OVF. These data indicate that EBN possesses anti-aging and neurodegenerative effects. In addition, it was observed that the EBN extract protects dopaminergic neurons from six hydroxdopoamine-induced degeneration. Reduced apoptosis in human neuroblastoma SH-SY5Y cells results in enhanced cell survival (Yew et al. 2014). EBN might be a viable treatment for neurodegenerative diseases such as Alzheimer's disease (AD) and Parkinson's disease that are aggravated by oxidative stress.

6.3.5 CE Status of EBN

Increasing age is among the most prominent risk factors for Alzheimer's disease dementia. Cognitive decline may be induced by a reduction in cerebral blood flow (CBF), which may initiate a cascade of inflammation and oxidative stress. Recent medical theory has emphasized the importance of natural antioxidants as a nutrient for protecting the brain against physiological changes that result in aging or neurological disease. In one of our studies, bilateral occlusion of the common carotid arteries (2VO) was employed to induce a decrease in CBF in rats, simulating human brain CBF reduction with aging (Ismail et al. 2021). An irreversible and persistent blockage adds to pathophysiological modifications that increase oxidative stress and cause neuronal damage (Liu and Zhang 2012; Zhang et al. 2018). The CA1 region of the hippocampus is the most affected area of the brain, according per this research. As neurons degenerate, they contribute to substantial functional and morphological defects that are associated with the steady decline of memory and cognitive functions. Oral gavage was utilized to treat EBN for 8 weeks; it was delivered on the first postoperative day. This medicine may reduce neuronal cell damage resulting from CBF. Long-term carotid artery occlusion in the untreated group was associated with neuronal impairment and increased oxidative stress. An increase in the number of viable neuronal cells in the CA1 region of the hippocampus was indicative of a reduction in neuronal cell degeneration in 2VO-treated groups. In addition, oxidative stress generated by F2-isoprostane was reduced in the hippocampus compared to untreated groups. It has been established that EBN's antioxidant and anti-inflammatory properties have the potential to enhance cognitive functions, as indicated by its pharmacological intervention. Consuming foods with therapeutic characteristics may be effective for halting the course of Alzheimer's disease. In light of the fact that EBN has been used for centuries for medical and health-related purposes, our results indicate that it may be able to delay the development of Alzheimer's-related dementia when taken early in life. It may inhibit the aging of neurons when taken as a supplement.

6.4 Discussion

We defined cognitive enhancement for the sake of this chapter as an increase in proficiency with fundamental aspects of cognitive processes necessary for everyday life. In healthy individuals, treatment with CEs has been shown to improve a key indication of cognitive processes for daily behavior. The same holds true for individuals with pre-existing cognitive impairment; they will need to exhibit substantial gains in brainpower for the therapy to be considered successful. It is generally agreed that attention and memory are two of the most important mental capabilities. The modifications must be evaluated using objective tests that are reliable, valid, and sensitive in measuring the important aspects of function. Finally, randomized, double-blind studies with a placebo control should be the norm in cognitive research just as they are in psychopharmacology today.

The synthesis of new proteins in brain cells and the rearrangement of existing ones among dendritic processes in three dimensions are both hypothesized to play a part in the consolidation of memories. We now know the electrophysiological basis for conditioned cellular behavior via the discovery of long-term potentiation and the identification of some of the cellular signaling pathways involved. Despite this, and despite the vast amounts of data we have amassed, we have not come any closer to achieving our long-sought aim of improved mental capacity. However, similar results may be achieved with the help of pharmaceuticals and natural items, as is often claimed in the media. Some dietary brain boosters are usually accepted as safe, but the idea of greatly enhancing cognitive capacity or memory and learning in otherwise cognitively normal adults by regular pharmacological treatments is sometimes considered as morally dubious. Most rigorous scientific study has so far been directed to Alzheimer's disease, with an emphasis on reversing cognitive decline in dementia. Drugs that have shown promise in preventing or treating dementia in people or animal models often show the same promise in healthy control participants. Because of this, medicines and their natural product analogues will probably likely be employed to enhance cognitive performance in a variety of contexts.

It comes as no surprise that researchers are looking at natural chemicals that might boost human intelligence. It would take a couple more books just to cover the ground that human studies occupy. Although the outcomes of peer-reviewed articles in human cognitive psychopharmacology need rigorous trials before publication, the "natural product" sector is generally characterized as a "soft science." Many claims are supported by anecdotes, uncontrolled research, or experiments on animals. Despite its usefulness in selecting study participants, this information does not constitute "core evidence" for a claim that a chemical improves cognitive performance.

This chapter emphasizes that EBN has the potential to enhance cognitive functioning (Ismail et al. 2021). There are several methods that have been recommended to help with fundamental skills like remembering and paying attention. The most pressing issues facing EBN right now are determining which populations are most suited for development, how much progress is possible in which areas of function, how long this may take, and which areas of function can be improved upon. Improvements in mental capacity are more difficult to discern than declines. The window of opportunity to increase function in healthy people is much narrower than the window of opportunity to lower it. As we have seen in the preceding pages, the average person's quality of life may improve by no more than 15%, whereas severely impaired people can enjoy a 100% increase. As a result, the investigations on EBN's potential as CEs must do all reasonably necessary to maximize the intensity of the signal.

EBN has been shown to improve mental performance. Since the mechanism of action and chemical structure of EBN are well understood, the drug will be offered in standardized formulations in which the quantities and ratios of various constituents are guaranteed to be present within acceptable ranges. While the material shows promise, very little study has been done on it thus far. These investigations cannot be replicated using identical formulations due to the lack of standardization of the known active components in EBN. No conclusive evidence of effectiveness, especially that related to cognitive function after long-term EBN usage in humans, exists at this time. The lack of comprehensive literature in this area may contribute to the difficulty of elucidating this topic.

6.5 Concluding Remarks and Suggestions

EBN is used extensively all around the globe because to its many practical applications. The nutritional and pharmacological qualities of the EBN product must be taken into account throughout its development. Incorporating scientific research data into EBN's value may increase customer trust in the company's products and services. Few studies have shown EBN's effect on mental performance. We recommend doing a behavioral experiment to determine whether or not EBN supplementation enhances cognitive function in an AD rat model. Anti-inflammatory indicators should be tested to learn more about the EBN's protective effects. Studies in the future should use the same animal model to analyze gene and protein expression in both the hippocampus and cerebral cortex.

References

Albishtue AA, Yimer N, Zakaria MZA, Haron AW, Babji AS, Abubakar AA, Almhanawi BH (2019) Effects of EBN on embryo implantation, plasma concentrations of reproductive hormones, and uterine expressions of genes of PCNA, steroids, growth factors and their receptors in rats. *Theriogenology*. https://doi.org/10.1016/j.theriogenology.2018.12.026

Alcohol and Drug Foundation. Nootropics (2020). Available at: https://adf.org.au/drug-facts/cognitive-enhancers/ (Accessed October 2021).

Allen AL, Strand NK (2015) Cognitive enhancement and beyond: recommendations from the Bioethics Commission. *Trends Cogn Sci 19*(10):549–551. https://doi.org/10.1016/j.tics.2015.08.001

Arvanitakis Z, Shah RC, Bennett DA (2019) Diagnosis and management of dementia: review. *JAMA 322*(16):1589–1599. https://doi.org/10.1001/jama.2019.4782

Aston-Jones G, Cohen JD (2005) An integrative theory of locus coeruleus-norepinephrine function: adaptive gain and optimal performance. *Ann Rev Neurosci 28*(1):403–450. https://doi.org/10.1146/annurev.neuro.28.061604.135709

Aswir AR, Wan Nazaimoon WM (2011) Effect of edible bird's nest on cell proliferation and tumor necrosis factor- alpha (TNF-α) release in vitro. *Int Food Res J*

Babji AS, Etty Syarmila IK, Nur'Aliah D, Nurul Nadia M, Hadi Akbar D, Norrakiah AS, Ghassem M, Najafian L, Salma MY (2018) Assessment on bioactive components of hydrolysed edible bird nest. Int Food Res J 25(5):1936–1941

Barceló B, Gomila I, Rotolo MC, Marchei E, Kyriakou C, Pichini S et al (2017) Intoxication caused by new psychostimulants: analytical methods to disclose acute and chronic use of benzofurans and ethylphenidate. *Int J Legal Med 131*(6):1543–1553. https://doi.org/10.1007/s00414-017-1648-9

Barfejani AH, Jafarvand M, Seyedsaadat SM, Rasekhi RT (2020) Donepezil in the treatment of ischemic stroke: Review and future perspective. *Life Sci 263*:118575. https://doi.org/10.1016/j.lfs.2020.118575

Bergström LS, Lynöe N (2008) Enhancing concentration, mood and memory in healthy individuals: An empirical study of attitudes among general practitioners and the general population. *Scand J Public Health 36*(5):532–537. https://doi.org/10.1177/1403494807087558

Bostrom N, Sandberg A (2009) Cognitive enhancement: methods, ethics, regulatory challenges. *Sci Eng Ethics 15*(3):311–341. https://doi.org/10.1007/s11948-009-9142-5

Bublitz JC, Merkel R (2012) Crimes against minds: on mental manipulations, harms and a human right to mental self-determination. *Crim Law Philos 8*(1):51–77. https://doi.org/10.1007/s11572-012-9172-y

Careena S, Sani D, Tan SN, Lim CW, Hassan S, Norhafizah M, Kirby BP, Ideris A, Stanslas J, Basri HB, Lim CTS (2018) Effect of edible bird's nest extract on lipopolysaccharide-induced impairment of learning and memory in wistar rats. Evid Based Complement Alternat Med 2018:9318789. https://doi.org/10.1155/2018/9318789. eCollection 2018

Carson SH, Fau PJB, Higgins DM, Higgins DM (2003) Decreased latent inhibition is associated with increased creative achievement in high-functioning individuals. J Pers Soc Psychol 85(3):499–506

Caviola L, Faber NS (2015) Pills or push-ups? Effectiveness and public perception of pharmacological and non-pharmacological cognitive enhancement. *Front Psychol 6*:1852. https://doi.org/10.3389/fpsyg.2015.01852

Caviola L, Mannino A, Savulescu J, Faulmuller N (2014) Cognitive biases can affect moral intuitions about cognitive enhancement. *Front Syst Neurosci 8*:195. https://doi.org/10.3389/fnsys.2014.00195

Caye A, Swanson JM, Coghill D, Rohde LA (2018) Treatment strategies for ADHD: an evidence-based guide to select optimal treatment. *Mol Psychiatry 24*(3):390–408. https://doi.org/10.1038/s41380-018-0116-3

Chua KH, Lee TH, Kamini N, Nor Hamdan MY, Lee CT, Eddie TTT, Chau LS, Zukefl SN (2016) A comprehensive review on edible bird nests and swiftlet farming. J Integr Med 14(6):415–428

Chua YG, Bloodworth BC, Leong LP, Li SFY (2014) Metabolite profiling of edible bird's nest using gas chromatography/mass spectrometry and liquid chromatography/mass spectrometry. *Rapid Commun Mass Spectrom 28*(12):1387–1400. https://doi.org/10.1002/rcm.6914

Chua YG, Chan SH, Bloodworth BC, Li SFY, Leong LP (2015) Identification of edible birds nest with amino acid and monosaccharide analysis. *J Agric Food Chem 63*(1):279–289. https://doi.org/10.1021/jf503157n

Chuah LYM, Chong DL, Chen AK, Rekshan WR 3rd, Tan J-C, Zheng H, Chee MWL (2009) Donepezil improves episodic memory in young individuals vulnerable to the effects of sleep deprivation. *Sleep 32*(8):999–1010. https://doi.org/10.1093/sleep/32.8.999

Clatworthy PL, Lewis SJG, Brichard L, Hong YT, Izquierdo D, Clark L et al (2009) Dopamine release in dissociable striatal subregions predicts the different effects of oral methylphenidate on reversal learning and spatial working memory. *J Neurosci 29*(15):4690–4696. https://doi.org/10.1523/JNEUROSCI.3266-08.2009

Cools R, Barker RA, Sahakian BJ, Robbins TW (2001) Enhanced or impaired cognitive function in Parkinson's disease as a function of dopaminergic medication and task demands. *Cerebral Cortex 11*(12):1136–1143. https://doi.org/10.1093/cercor/11.12.1136

Cools R, Roberts AC, Robbins TW (2008) Serotoninergic regulation of emotional and behavioural control processes. *Trends Cogn Sci 12*(1):31–40. https://doi.org/10.1016/j.tics.2007.10.011

Dagher A, Robbins TW (2009) Personality, addiction, dopamine: insights from Parkinson's disease. *Neuron 61*(4):502–510. https://doi.org/10.1016/j.neuron.2009.01.031

Dayan P, Huys QJM (2009) Serotonin in affective control. *Ann Rev Neurosci 32*(1):95–126. https://doi.org/10.1146/annurev.neuro.051508.135607

Denzer C, Denzer F, Lennerz BS, Vollbach H, Lustig RH, Wabitsch M (2019) Treatment of hypothalamic obesity with dextroamphetamine: a case series. *Obesity Facts 12*(1):91–102. https://doi.org/10.1159/000495851

Dresler M, Sandberg A, Bublitz C, Ohla K, Trenado C, Mroczko-Wasowicz A et al (2019) Hacking the brain: dimensions of cognitive enhancement. *ACS Chem Neurosci 10*(3):1137–1148. https://doi.org/10.1021/acschemneuro.8b00571

Earp BD, Sandberg A, Kahane G, Savulescu J (2014) When is diminishment a form of enhancement? Rethinking the enhancement debate in biomedical ethics. *Front Syst Neurosci 8*:12–12. https://doi.org/10.3389/fnsys.2014.00012

Edgar CJ, Pace-Schott EF, Wesnes KA (2009) Approaches to measuring the effects of wake-promoting drugs: a focus on cognitive function. *Human Psychopharmacol 24*(5):371–389. https://doi.org/10.1002/hup.1034

Farah MJ, Illes J, Cook-Deegan R, Gardner H, Kandel E, King P et al (2004) Neurocognitive enhancement: what can we do and what should we do? *Nat Rev Neurosci 5*(5):421–425. https://doi.org/10.1038/nrn1390

Faulmuller N, Maslen H, de Sio FS (2013) The indirect psychological costs of cognitive enhancement. *Am J Bioeth 13*(7):45–47. https://doi.org/10.1080/15265161.2013.794880

FitzGerald DB, Crucian GP, Mielke JB, Shenal BV, Burks D, Womack KB et al (2008) Effects of donepezil on verbal memory after semantic processing in healthy older adults. *Cogn Behav Neurol 21*(2):57–64. https://doi.org/10.1097/wnn.0b013e3181799df1

Floresco SB, Magyar O (2006) Mesocortical dopamine modulation of executive functions: beyond working memory. *Psychopharmacology 188*(4):567–585. https://doi.org/10.1007/s00213-006-0404-5

Ghassem M, Arihara K, Mohammadi S, Sani NA, Babji AS (2017) Identification of two novel antioxidant peptides from edible bird's nest (Aerodramus fuciphagus) protein hydrolysates. *Food Function 8*(5):2046–2052. https://doi.org/10.1039/c6fo01615d

Glannon W (2015) Neuroethics. In *Oxford Handbooks Online*: Oxford University Press.

Greely H, Sahakian B, Harris J, Kessler RC, Gazzaniga M, Campbell P, Farah MJ (2008) Towards responsible use of cognitive-enhancing drugs by the healthy. *Nature 456*(7223):702–705. https://doi.org/10.1038/456702a

Grossberg GT, Tong G, Burke AD, Tariot PN (2019) Present algorithms and future treatments for Alzheimer's disease. *J Alzheimer's Dis 67*:1157–1171. https://doi.org/10.3233/JAD-180903

Guo CT, Takahashi T, Bukawa W, Takahashi N, Yagi H, Kato K, Hidari KIPJ, Miyamoto D, Suzuki T, Suzuki Y (2006) Edible bird's nest extract inhibits influenza virus infection. *Antiviral Res 70*(3):140–146. https://doi.org/10.1016/j.antiviral.2006.02.005

Haghani A, Mehrbod P, Safi N, Aminuddin NA, Bahadoran A, Omar AR, Ideris A (2016) In vitro and in vivo mechanism of immunomodulatory and antiviral activity of Edible Bird's Nest (EBN) against influenza A virus (IAV) infection. *J Ethnopharmacol 185*:327–340. https://doi.org/10.1016/j.jep.2016.03.020

Haghani A, Mehrbod P, Safi N, Kadir F, Aini A, Omar AR, Ideris A (2017) Edible bird's nest modulate intracellular molecular pathways of influenza A virus infected cells. BMC Complement Altern Med 17:1. https://doi.org/10.1186/s12906-016-1498-x

Hall WD, Lucke JC (2010) The enhancement use of neuropharmaceuticals: more scepticism and caution needed. *Addiction 105*(12):2041–2043. https://doi.org/10.1111/j.1360-0443.2010.03211.x

Harvey PD, Bowie CR (2012) Cognitive enhancement in schizophrenia: pharmacological and cognitive remediation approaches. *Psychiatr Clin North Am 35*(3):683–698. https://doi.org/10.1016/j.psc.2012.06.008

Hauskeller M (2014) Better humans? In: Routledge.

Hobbs JJ (2004) Prolems in harvest of Edible birds'nest in Sarwak and Sabah, Malaysia&Borneo. Biodivers Conserv 13:2209–2226

Hou Z, He P, Imam MU, Qi J, Tang S, Song C, Ismail M (2017) Edible bird's nest prevents menopause-related memory and cognitive decline in rats via increased hippocampal sirtuin-1 expression. *Oxid Med Cell Longev 1-8*. https://doi.org/10.1155/2017/7205082

Hou Z, Imam MU, Ismail M, Ooi DJ, Ideris A, Mahmud R (2015) Nutrigenomic effects of edible bird's nest on insulin signaling in ovariectomized rats. *Drug Design Dev Ther 9*:4115–4125. https://doi.org/10.2147/DDDT.S80743

Hu Q, Li G, Yao H, He S, Li H, Liu S, Wu Y, Lai X (2016) Edible bird's nest enhances antioxidant capacity and increases lifespan in Drosophila Melanogaster. Cell Mol Biol (Noisyle-grand) 62(4):116–122

Husain M, Mehta MA (2011) Cognitive enhancement by drugs in health and disease. *Trends Cogn Sci 15*(1):28–36. https://doi.org/10.1016/j.tics.2010.11.002

Ibrahim HR, Hoq MI, Aoki T (2007) Ovotransferrin possesses SODlike superoxide anion scavenging activity that is promoted by copper and manganese binding. Int J Biol Macromol 2007(41):631–640

Ismail R, Hui C, Affandi K, Allallam B, Mohamed WM, Noor M (2021) Neuroprotective effect of edible bird's nest in chronic cerebral hypoperfusion induced neurodegeneration in rats. Neuroimmunol Neuroinflamm. https://doi.org/10.20517/2347-8659.2020.63

Iuculano T, Cohen Kadosh R (2013) The mental cost of cognitive enhancement. *J Neurosci 33*(10):4482–4486. https://doi.org/10.1523/JNEUROSCI.4927-12.2013

de Jongh R, Bolt I, Schermer M, Olivier B (2008) Botox for the brain: enhancement of cognition, mood and pro-social behavior and blunting of unwanted memories. *Neurosci Biobehav Rev 32*(4):760–776. https://doi.org/10.1016/j.neubiorev.2007.12.001

Kabir MT, Uddin MS, Mamun AA, Jeandet P, Aleya L, Mansouri RA et al (2020) Combination drug therapy for the management of Alzheimer's disease. *Int J Mol Sci 21*(9):3272. https://doi.org/10.3390/ijms21093272

Kathan RH, Weeks DI (1969) Structure studies of collocalia mucoid. I. Carbohydrate and amino acid composition. *Arch Biochem Biophys 134*(2):572–576. https://doi.org/10.1016/0003-9861(69)90319-1

Khalida SKA, Rashedb AA, Azizc SA, Ahmada H (2019) Effects of sialic acid from edible bird nest on cell viability associated with brain cognitive performance in mice. World J Trad Chin Med 5(4):214

Kokiko ON, Hamm RJ (2007) A review of pharmacological treatments used in experimental models of traumatic brain injury. *Brain Inj 21*(3):259–274. https://doi.org/10.1080/02699050701209964

Kong YC, Keung WM, Yip TT, Ko KM, Tsao SW, Ng MH (1987) Evidence that epidermal growth factor is present in swiftlet's (Collocalia) nest. *Comp Biochem Physiol Part B: Biochem And 87*(2):221–226. https://doi.org/10.1016/0305-0491(87)90133-7

Lees J, Michalopoulou PG, Lewis SW, Preston S, Bamford C, Collier T et al (2017) Modafinil and cognitive enhancement in schizophrenia and healthy volunteers: the effects of test battery in a randomised controlled trial. *Psychol Med 47*(13):2358–2368. https://doi.org/10.1017/S0033291717000885

Liakoni E, Schaub MP, Maier LJ, Glauser G-V, Liechti ME (2015) The use of prescription drugs, recreational drugs, and "soft enhancers" for cognitive enhancement among Swiss Secondary School Students. *PLoS One 10*(10):e0141289–e0141289. https://doi.org/10.1371/journal.pone.0141289

Liu H, Zhang J (2012) Cerebral hypoperfusion and cognitive impairment: The pathogenic role of vascular oxidative stress. *Int J Neurosci 122*(9):494–499. https://doi.org/10.3109/00207454.2012.686543

Luciana M, Burgund ED, Berman M, Hanson KL (2001) Effects of tryptophan loading on verbal, spatial and affective working memory functions in healthy adults. *J Psychopharmacol 15*(4):219–230. https://doi.org/10.1177/026988110101500410

Lucke J, Partridge B (2012) Towards a smart population: a public health framework for cognitive enhancement. *Neuroethics 6*(2):419–427. https://doi.org/10.1007/s12152-012-9167-3

Ma F, Liu D (2012) Sketch of the edible bird's nest and its important bioactivities. Food Res Int 48:559–567

Macer D (2012) Ethical consequences of the positive views of enhancement in Asia. *Health Care Analysis 20*(4):385–397. https://doi.org/10.1007/s10728-012-0230-3

Macron MF. (2005) Characterization of edible bird's nest the "caviar of East". food research international;1125-1134 .

Mahaq O, Rameli MAP, Jaoi Edward M, Mohd Hanafi N, Abdul Aziz S, Abu Hassim H, Mohd Noor MH, Ahmad H (2020) The effects of dietary edible bird nest supplementation on learning and memory functions of multigenerational mice. *Brain Behav 10*(11):e01817. https://doi.org/10.1002/brb3.1817

Marchei E, Palmi I, Pichini S, Pacifici R, Anton Airaldi I-R, Costa Orvay JA et al (2016) Análisis segmentario del pelo para detectar la exposición crónica a drogas psicoactivas. *Adicciones 28*(3):158. https://doi.org/10.20882/adicciones.825

Matsukawa N, Matsumoto, Bukawa W, Chiji H, Nakayama K, Hara H, Tsukahara T (2011) Improvement of bone strength and dermal thickness due to dietary edible bird's nest extract in ovariectomized rats. Biosci Biotechnol Biochem 75(3):590–592

Medaglia JD, Yaden DB, Helion C, Haslam M (2019) Moral attitudes and willingness to enhance and repair cognition with brain stimulation. *Brain Stimul 12*(1):44–53. https://doi.org/10.1016/j.brs.2018.09.014

Mehta MA, Swainson R, Ogilvie AD, Sahakian B, Robbins TW (2001) Improved short-term spatial memory but impaired reversal learning following the dopamine D2 agonist bromocriptine in human volunteers. *Psychopharmacology 159*(1):10–20. https://doi.org/10.1007/s002130100851

Metzinger T, & Hildt E (2011) Cognitive enhancement. In *Oxford Handbooks Online*: Oxford University Press.

Müller U, Steffenhagen N, Regenthal R, Bublak P (2004) Effects of modafinil on working memory processes in humans. *Psychopharmacology 177*(1):161–169. https://doi.org/10.1007/s00213-004-1926-3

Naismith SL, Lewis SJG, & Rogers NL (2011) Sleep–wake changes and cognition in neurodegenerative disease. In *Progress in Brain Research* (pp. 21–52): Elsevier.

Olivares D, Deshpande KV, Shi Y, Lahiri DK, Greig HN, Rogers JT, Huang X (2012) N-Methyl D-Aspartate (NMDA) receptor antagonists and memantine treatment for Alzheimer's disease, vascular dementia and Parkinson's disease. *Curr Alzheimer Res 9*(6):746–758. https://doi.org/10.2174/156720512801322564

Pavarini G, McKeown A, Singh I (2018) Smarter than thou, holier than thou: the dynamic interplay between cognitive and moral enhancement. *Front Pharmacol 9*:1189. https://doi.org/10.3389/fphar.2018.01189

Pieramico V, Esposito R, Cesinaro S, Frazzini V, Sensi SL (2014) Effects of non-pharmacological or pharmacological interventions on cognition and brain plasticity of aging individuals. *Front Syst Neurosci 8*:153–153. https://doi.org/10.3389/fnsys.2014.00153

Posner J, Polanczyk GV, Sonuga-Barke E (2020) Attention-deficit hyperactivity disorder. *Lancet (London, England) 395*(10222):450–462. https://doi.org/10.1016/S0140-6736(19)33004-1

Pozsgay V, Jennings H, Kasper DL (1987) 4,8-anhyro-N acetylneuraminic acid. Isolation from edible bird's nest and structure deremination. Eur J Biochem 162(2):445–450. https://doi.org/10.1111/j.1432-1033.1987.tb10622.x

Quek MC, Chin NL, Yusof YA, Law CL, Tan SW (2018) Characterization of edible bird's nest of different production, species and geographical origins using nutritional composition, physicochemical properties and antioxidant activities. *Food Res Int 109*:35–43. https://doi.org/10.1016/j.foodres.2018.03.078

Racine E, Sattler S, Boehlen W (2021) Cognitive enhancement: unanswered questions about human psychology and social behavior. *Sci Eng Ethics 27*(2):19. https://doi.org/10.1007/s11948-021-00294-w

Ram SS, Russell B, Kirkpatrick C, Stewart K, Scahill S, Henning M et al (2020) Professionals' attitudes towards the use of cognitive enhancers in academic settings. *PLoS One 15*(11):e0241968. https://doi.org/10.1371/journal.pone.0241968

Repantis D, Laisney O, Heuser I (2010) Acetylcholinesterase inhibitors and memantine for neuroenhancement in healthy individuals: a systematic review. *Pharmacol Res 61*(6):473–481. https://doi.org/10.1016/j.phrs.2010.02.009

Riis J, Simmons JP, Goodwin GP (2008) Preferences for enhancement pharmaceuticals: the reluctance to enhance fundamental traits. *J Consumer Res 35*(3):495–508. https://doi.org/10.1086/588746

Robbins TW, Arnsten AFT (2009) The neuropsychopharmacology of fronto-executive function: monoaminergic modulation. *Ann Rev Neurosci 32*:267–287. https://doi.org/10.1146/annurev.neuro.051508.135535

Rousseau E, Michel PP, Hirsch EC (2013) The iron-binding protein lactoferrin protects vulnerable dopamine neurons from degeneration by preserving mitochondrial calcium homeostasis. Mol Pharmacol 2013(84):888–898

Sabini J, Monterosso J (2005) Judgments of the fairness of using performance enhancing drugs. *Ethics Behav 15*(1):81–94. https://doi.org/10.1207/s15327019eb1501_6

Sandberg A, Bostrom N (2006) Converging cognitive enhancements. *Ann NY Acad Sci 1093*(1):201–227. https://doi.org/10.1196/annals.1382.015

Sarter M, Parikh V, Howe WM (2009) Phasic acetylcholine release and the volume transmission hypothesis: time to move on. *Nat Rev Neurosci 10*(5):383–390. https://doi.org/10.1038/nrn2635

Scheffler RM, Fau BTT, Fulton BD, Fau FBD, Hinshaw SP, Fau HSP, Levine P, Fau LP, Stone S, Stone S (2009) Positive association between attention-deficit/ hyperactivity disorder medication use and academic achievement during elementary school. 123(5):1273–1279

Schelle KJ, Faulmuller N, Caviola L, Hewstone M (2014) Attitudes toward pharmacological cognitive enhancement-a review. *Front Syst Neurosci 8*:53. https://doi.org/10.3389/fnsys.2014.00053
Schleim S (2014) Whose well-being? Common conceptions and misconceptions in the enhancement debate. *Front Syst Neurosci 8*:148. https://doi.org/10.3389/fnsys.2014.00148
Schultz W (2002) Getting formal with dopamine and reward. *Neuron 36*(2):241–263. https://doi.org/10.1016/s0896-6273(02)00967-4
Seow EK, Ibrahim B, Muhammad SA, Lee LH, Cheng LH (2016) Differentiation between house and cave edible bird's nests by chemometric analysis of amino acid composition data. LWT Food Sci Technol 65:428–435
Singhammer J (2012) Age and gender specific variations in attitudes to performance enhancing drugs and methods. A cross-sectional study. *Sport Sci Rev 21*(5-6):29–48. https://doi.org/10.2478/v10237-012-0017-3
Swainson R, Rogers RD, Sahakian BJ, Summers BA, Polkey CE, Robbins TW (2000) Probabilistic learning and reversal deficits in patients with Parkinson's disease or frontal or temporal lobe lesions: possible adverse effects of dopaminergic medication. *Neuropsychologia 38*(5):596–612. https://doi.org/10.1016/s0028-3932(99)00103-7
Szelenberger R, Kostka J, Saluk-Bijak J, Miller E (2020) Pharmacological interventions and rehabilitation approach for enhancing brain self-repair and stroke recovery. *Curr Neuropharmacol 18*(1):51–64. https://doi.org/10.2174/1570159X17666190726104139
Szeto JYY, Lewis SJG (2016) Current treatment options for Alzheimer's disease and parkinson's disease dementia. *Curr Neuropharmacol 14*(4):326–338. https://doi.org/10.2174/1570159x14666151208112754
Tahami Monfared AA, Desai M, Hughes R, Lucherini S, Yi Y, Perry R (2020) Treatment options for dementia with lewy bodies: a network meta-analysis of randomised control trials. *Neurol Ther 9*(2):521–534. https://doi.org/10.1007/s40120-020-00198-0
Veroniki AA, Straus SE, Ashoor HM, Hamid JS, Hemmelgarn BR, Holroyd-Leduc J et al (2016) Comparative safety and effectiveness of cognitive enhancers for Alzheimer's dementia: protocol for a systematic review and individual patient data network meta-analysis. *BMJ Open 6*(1):e010251–e010251. https://doi.org/10.1136/bmjopen-2015-010251
Voon V, Fernagut P-O, Wickens J, Baunez C, Rodriguez M, Pavon N et al (2009) Chronic dopaminergic stimulation in Parkinson's disease: from dyskinesias to impulse control disorders. *Lancet Neurol 8*(12):1140–1149. https://doi.org/10.1016/s1474-4422(09)70287-x
Wang B, Yu B, Karim M, Hu H, Sun Y, Mc Greevy P, Brand-Miller J (2007) Dietary sialic acid supplementation improves learning and memory in piglets. Am J Clin Nutr 85(2):561–569. https://doi.org/10.1093/ajcn/85.2.561
Wang B (2009) Sialic acid is an essential nutrient for brain development and cognition. *Annu Rev Nutr* 29:177–222. https://doi.org/10.1146/annurev.nutr.28.061807.155515
Wang B (2012) Molecular mechanism underlying sialic acid as an essential nutrient for brain development and cognition. *Adv Nutr 3*:3. https://doi.org/10.3945/an.112.001875
Wang CC (1921) The composition of Chinese edible bird's nest and the nature of their proteins. J Biol Chem 49:429–439
Weintraub D, Koester J, Potenza MN, Siderowf AD, Stacy M, Voon V et al (2010) Impulse control disorders in Parkinson disease. *Arch Neurol 67*:5. https://doi.org/10.1001/archneurol.2010.65
Weintraub D, Siderowf AD, Potenza MN, Goveas J, Morales KH, Duda JE et al (2006) Association of dopamine agonist use with impulse control disorders in Parkinson disease. *Arch Neurol 63*(7):969–973. https://doi.org/10.1001/archneur.63.7.969
Wesnes KA, McKeith I, Edgar C, Emre M, Lane R (2005) Benefits of rivastigmine on attention in dementia associated with Parkinson disease. *Neurology 65*(10):1654–1656. https://doi.org/10.1212/01.wnl.0000184517.69816.e9
Wezenberg E, Verkes RJ, Sabbe BGC, Ruigt GSF, Hulstijn W (2005) Modulation of memory and visuospatial processes by biperiden and rivastigmine in elderly healthy subjects. *Psychopharmacology 181*(3):582–594. https://doi.org/10.1007/s00213-005-0083-7

Wilms W, Woźniak-Karczewska M, Corvini PFX, Chrzanowski Ł (2019) Nootropic drugs: methylphenidate, modafinil and piracetam—Population use trends, occurrence in the environment, ecotoxicity and removal methods—A review. *Chemosphere 233*:771–785. https://doi.org/10.1016/j.chemosphere.2019.06.016

Wise RA (2004) Dopamine, learning and motivation. *Nat Rev Neurosci 5*(6):483–494. https://doi.org/10.1038/nrn1406

Wong C-F, Chan GK-L, Zhang M-L, Yao P, Lin H-Q, Dong TT-X, Li G, Lai X-P, Tsim KW-K (2017) Characterization of edible bird's nest by peptide fingerprinting with principal component analysis. Food Qual Safe 1:83–92

Wong RSY (2013) Edible bird's nest: food or medicine? *Chin J Integr Med 19*(9):643–649. https://doi.org/10.1007/s11655-013-1563-y

Wong ZCF, Chan GKL, Wu KQY, Poon KKM, Chen Y, Dong TTX, Tsim KWK (2018) Complete digestion of edible bird's nest releases free N-acetylneuraminic acid and small peptides: an efficient method to improve functional properties. *Food Funct 9*(10):5139–5149. https://doi.org/10.1039/c8fo00991k

Xie Y, Zeng H, Huang Z, Xu H, Fan Q, Zhang Y, Zheng B (2018) Effect of maternal administration of edible bird's nest on the learning and memory abilities of suckling offspring in mice. *Neural Plasticity 2018*. https://doi.org/10.1155/2018/7697261

Xin WH, Kasim ZM, Babji AS, Ismail NH, Muhammad NN (2014). Proximate analysis and amino acid composition in selected edible bird's nest. The 16th Food Innovation Asia Conference, Bangkok, Thailand.

Yew MY, Koh RY, Chye SM, Othman I, Ng KY (2014) Edible bird's nest ameliorates oxidative stress-induced apoptosis in SH-SY5Y human neuroblastoma cells. *BMC Complement Alternat Med 14*:391. https://doi.org/10.1186/1472-6882-14-391

Yida Z, Imam MU, Ismail M, Hou Z, Abdullah MA, Ideris A, Ismail N (2015a) Edible Bird's Nest attenuates high fat diet-induced oxidative stress and inflammation via regulation of hepatic antioxidant and inflammatory genes. *BMC Complement Alternat Med 15*:1. https://doi.org/10.1186/s12906-015-0843-9

Yida Z, Imam MU, Ismail M, Ismail N, Hou Z (2015b) Edible bird's nest attenuates procoagulation effects of high-fat diet in rats. *Drug Design Dev Ther 9*:3951–3959. https://doi.org/10.2147/DDDT.S87772

Yida Z, Imam UU, Ismail M (2014) In vitro bioaccessibility and antioxidant properties of edible bird's nest following simulated human gastro-intestinal digestion. *BMC Complement Alternat Med 14*:1. https://doi.org/10.1186/1472-6882-14-468

Zaami S, Tagliabracci A, Berretta P, Busardò FP, Marinelli E (2020) Use of methylphenidate analogues as cognitive enhancers: the prelude to cosmetic neurology and an ethical issue. *Front Psychiatry 10*:1006–1006. https://doi.org/10.3389/fpsyt.2019.01006

Zhang D, Xiao Y, Lv P, Teng Z, Dong Y, Qi Q, Liu Z (2018) Edaravone attenuates oxidative stress induced by chronic cerebral hypoperfusion injury: role of ERK/Nrf2/HO-1 signaling pathway. *Neurol Res 40*(1):1–10. https://doi.org/10.1080/01616412.2017.1376457

Zhao R, Li G, Kong XJ, Huang XY, Li W, Zeng YY, Lai XP (2016) The improvement effects of edible bird's nest on proliferation and activation of B lymphocyte and its antagonistic effects on immunosuppression induced by cyclophosphamide. *Drug Design Dev Ther 10*:371–381. https://doi.org/10.2147/DDDT.S88193

Zhiping H, Imam MU, Ismail M, Ismail N, Yida Z, Ideris A et al (2015) Effects of edible bird's nest on hippocampal and cortical neurodegeneration in ovariectomized rats. *Food Funct 6*(5):6.:1701-11. https://doi.org/10.1039/c5fo00226e

Chapter 7
Trace Elements and Epilepsy

Abdelaati El Khiat, Driss Ait Ali, Bilal El-Mansoury, Youssef Ait Hamdan, Brahim El Houate, Mohamed El Koutbi, Lahcen Tamegart, Halima Gamrani, and Najib Kissani

Abstract Epilepsy is a chronic neurological disease, resulting in significant mortality worldwide. Several experimental and clinical works have suggested that the normal brain function depends on the homeostasis of trace elements. However, taking antiepileptics can also modify the homeostasis of trace elements. The objective of this chapter was to carry out a bibliographical synthesis to establish the link between epilepsy and the alteration of the concentration of certain trace elements, in particular, zinc, copper, magnesium, and selenium in the blood and/or in the brain of epileptic patients. The results found during this survey reinforce the role of trace

A. El Khiat (✉)
Laboratory of Clinical and Experimental Neurosciences and Environment, Faculty of Medicine and Pharmacy, Cadi Ayyad University, Marrakech, Morocco

Biological and health sciences team, Higher Institute of Nursing Professions and Health Techniques, Ministry of Health, Ouarzazate, Morocco

Interdisciplinary Laboratory in Bio-Resources, Environment and Materials, Higher Normal School, Marrakech, Morocco

D. Ait Ali · B. El Houate · M. El Koutbi
Biological and health sciences team, Higher Institute of Nursing Professions and Health Techniques, Ministry of Health, Ouarzazate, Morocco

B. El-Mansoury
Faculty of Sciences, Department of Biology, Chouaib Doukkali University, El Jadida, Morocco

Y. Ait Hamdan
Interdisciplinary Laboratory in Bio-Resources, Environment and Materials, Higher Normal School, Marrakech, Morocco

L. Tamegart
Laboratory of Clinical and Experimental Neurosciences and Environment, Faculty of Medicine and Pharmacy, Cadi Ayyad University, Marrakech, Morocco

Department of Biology, Faculty of Science, Abdelmalek Essaadi University, Tetouan, Morocco

H. Gamrani · N. Kissani
Laboratory of Clinical and Experimental Neurosciences and Environment, Faculty of Medicine and Pharmacy, Cadi Ayyad University, Marrakech, Morocco

W. Mohamed, R. Sandhir (eds.), *Trace Elements in Brain Health and Diseases*, Nutritional Neurosciences, https://doi.org/10.1007/978-981-99-1513-2_7

elements, whether directly or indirectly, in the pathogenesis of this disease, as well as its resistance to antiepileptic treatments.

Keywords Epilepsy · Seizures · Trace elements · Antiepileptic treatment

7.1 Introduction

Epilepsy, the most common serious neurological disease, represents 0.6% of the global burden of disease (Kissani et al. 2021), with a population of about 50 million people with epilepsy worldwide, 10 million of them in Africa. In Morocco, epilepsy affects around 374,000 people (El Khiat et al. 2021). The psychological, physical, and social burden of this disease is now recognized (Scott et al. 2001). Epilepsy is a persistent neurological disorder caused by a variety of factors. Despite substantial research, epilepsy development (epileptogenesis) is still poorly understood. The exact mechanisms behind the genesis and evolution of epilepsy are likely complex and have yet to be explored (Doboszewska et al. 2019).This may explain why there is currently no effective treatment for the disease and the available drugs tend only to stop or lower the frequency and intensity of seizures regardless of the epileptic underlying condition and disease evolution (Doboszewska et al. 2019). As such, exploring the pathophysiology of epilepsy is therefore critical for identifying new therapeutic targets and developing prevention strategies with favorable outcomes (Doboszewska et al. 2019).

Many trace elements such as zinc (Zn), copper (Cu), magnesium (Mg), calcium (Ca), iron (Ir), and selenium (Se) are recognized to have an essential position in the functioning of the brain, as well as the development and prevention of neurological illnesses including epilepsy (Doboszewska et al. 2019; Doretto et al. 2002; Khan 2016; Prakash et al. 2015; D. K. V. Prasad et al. 2014; Saad et al. 2014; Wojciak et al. 2013). As already identified, the mechanism of seizure is due to the imbalance of neuronal excitation and inhibition. These components are implicated in a variety of metabolic activities, including those in the human brain (Chwiej et al. 2008). The literature on the subject shows that epilepsy modifies the distribution of trace elements in tissues. Our aim, in this chapter, is to carry out investigations to establish the link between epilepsy and these trace elements especially zinc, copper, magnesium, and selenium.

7.2 Trace Elements Alterations in the Brain and Blood of Patients with Epilepsy

7.2.1 Trace Element Distribution in the Human Brain

The link between essential trace elements and epileptic seizures is unclear. Trace elements are necessary for human health and function (Lynes et al. 2007; Mehri 2020; Takeda 2004). However, these metals can be presented at concentrations that might alter normal physiologic activities. Hence, the balance of these trace elements is essential to maintain body homeostasis. In addition, these elements play a critical part in various biochemical processes, mostly as components of vitamins and enzymes, and are therefore responsible for a variety of metabolic activities in the brain (Chwiej et al. 2008). However, imbalances in the brain's trace elements might lead to brain disturbances. Several studies have reported alterations regarding trace elements in the serum and the brain tissues of patients with epilepsy. It has been argued that epileptic patients exhibit a variably of altered trace elements status, and it is believed that epileptic seizures highly modify the nervous tissue homeostasis. Hence, epileptic seizures might be a result of altered trace elements status (Chwiej et al. 2008). Indeed, trace elements excess and deficient conditions might lead to impaired neuronal maintenance that observed in neurological disorders especially via increasing oxidative stress. Moreover, these metals might be involved in the neurodegeneration in epilepsy (Chwiej et al. 2008) and therefore balanced trace elements in the brain are of great importance to prevent neuronal loss (Chwiej et al. 2008; Wandt et al. 2021). Growing evidence suggests that changes in metabolism and trace elements distribution within the brain might result in epileptic seizures (Carl et al. 1989; Hirate et al. 2002). Furthermore, these element levels might determine susceptibility to convulsions (Hirate et al. 2002). Besides, the information on how trace elements are distributed in the human brain and the correlation between composition with these elements' function are of great interest for identifying baseline regular values to be used as an indication of diseased brain tissue (Saiki et al. 2012). Hence, knowing trace element distribution in the brain of patients with epilepsy could be beneficial in finding effective treatment solutions. The metabolism of trace elements in the brain is influenced by the activity of seizures in epileptic patients and epilepsy experimental models (Carl et al. 1989; Hirate et al. 2002).

Researchers have demonstrated that disturbed zinc and copper homeostasis might be involved in susceptibility, development, as well as the termination of seizures, particularly in genetically determined epilepsy (Eissa et al. 2020). It should be noted that modifications in metal content do not have to happen throughout the brain or even within the entire anatomical systems, for instance, the hippocampus. Chwiej and coworkers, using X-ray fluorescence microscopy to examine the content and distribution of selected elements in the areas of the rat brains affected by pilocarpine-induced epilepsy, have detected that the Ca level was significantly higher in CA1 and CA3 regions of the hippocampus and in the cerebral cortex, whilst the copper level in the dentate gyrus and Zn in the CA3 region and the dentate gyrus were decreased (Chwiej et al. 2008). Subsequently, it is important to clarify whether increased vulnerability to seizures or epileptic neuronal activity affects the distribution of these trace elements in the brain (Hirate et al. 2002). The EL mouse (a genetically determined epilepsy animal model) is an appropriate model for studying the link between essential trace elements and seizure activity. Zinc concentration in the hippocampus dentate area of seized EL mice was found to be significantly decreased compared to control mice (ddY and CBA strains) (Fukahori et al. 1988). Furthermore, researchers used EL mice and ddY mice, which form the genetic basis for inbred EL mice, to study the distribution of trace elements in the brain, and discovered that Se concentration in the hippocampus was raised in the seized EL mice while Zn concentration in the cerebral cortex, cerebellum, and the rest of seized EL mice were lower (Hirate et al. 2002). Another study by Takeda et al. using EL mice shows that the induction of seizure lowers Zn concentrations in the brain (Takeda et al. 1999), suggesting that brain zinc homeostasis is linked to the pathophysiology of epileptic seizures. Moreover, the alteration of these trace elements homeostasis (zinc, cobalt, and selenium) in the brain might explain the susceptibility, development, or termination of seizures in EL mice (Hirate et al. 2002).

7.2.2 Trace Element Distribution in the Blood

The trace elements status in the serum of epilepsy patients was found to be altered. Indeed, studies have been conducted on the serum concentrations of trace elements such as copper, selenium, zinc, manganese, and iron in epileptic patients and animal models of epilepsy. The findings of these studies suggest a link between epilepsy and the low serum copper, manganese, and selenium levels (Kirkland et al. 2018; Toffa et al. 2018; Vitale et al. 2019; Yüzbaşioğlu et al. 2009).

7.2.2.1 Zinc and Epilepsy/Seizures

In many studies, human epileptic troubles have been linked to plasma Zn2+ deficiency. Nevertheless, the processes linking synoptically produced Zn2+ to seizure activity modulation are poorly understood (Khan 2016; Saad et al. 2014; Talat et al. 2015). Although there is some disagreement, studies have linked serum zinc levels to febrile seizure in children. When compared to the control groups, the average of serum zinc levels in the febrile seizure groups were considerably lower (Arul et al. 2020; Ganesh and Janakiraman 2008; Kheradmand et al. 2014; Lee and Kim 2012; Salehiomran and Mahzari 2013). In a Korean study, the serum zinc levels were evaluated in 288 children who had febrile seizures and 40 patients with afebrile seizures. The mean serum zinc levels were 60.5 ± 12.7 μg/dL in the febrile seizure group and 68.9 ± 14.5 μg/dL in the afebrile seizure group ($p < 0.001$) (Lee and Kim

2012).In the same order, Arul et al. (2020), in a study designed to determine the zinc serum in children under five with febrile seizures and febrile children who did not have seizures reveal two main significant results: The mean serum concentration of zinc in the cases and controls, respectively, was 83.8 33.1 μg/dL and 116.3 30.3 μg/dL ($p = 0.002$). Hypozincemia (serum zinc concentration less than 63 μg/dL. was detected in 12 (30%) cases and only in 2(5%) of controls with an odds ratio of 8:1(Arul et al. 2020). As for adult patients with epilepsy, in comparison to the control group, statistically significant low values of serum zinc levels were found (Farahani et al. 2013).

In addition, the relation between zinc deficiency, idiopathic intractable epilepsy, and genetic generalized epilepsy is also found to be significant when comparing cases and healthy children. Cases have significantly low serum zinc levels in comparison with healthy children (Kheradmand et al. 2014; D. K. V. Prasad et al. 2014; Seven et al. 2013). As zinc is used as a potent antioxidant in cellular interactions, alterations in his serum levels can trigger enzymatic changes, which can lead to neurological disorders (Farahani et al. 2013).Hypothetically, the inhibitory impact of zinc on GABA, an inhibitory neurotransmitter, is thought to be the mechanism through which zinc depletion enhances seizure activity (Doboszewska et al. 2019; Salah et al. 2014; Talat et al. 2015).Moreover, the NMDA receptor, which belongs to the glutamate family receptors, is activated by hypozincemia which may, therefore, be a key part in the origin of epileptic discharge (Namakin et al. 2016; Saad et al. 2014; Salah et al. 2014).

Interestingly, some studies reported the existence of a link between the serum zinc levels and the use of epileptic medications (Table 7.1). Serum zinc levels in epileptic patients taking valproate or carbamazepine were found to be lower than in controls. It is then possible that seizures are produced by high zinc levels or alternatively higher zinc levels are caused by seizure activity, and that pharmacologically reducing seizures help maintain normal zinc levels (Saghazadeh et al. 2015). These results are in accordance with those reported in the study carried out by Talat et al. (2015) demonstrating that the levels of serum zinc in epileptic children receiving treatment are lower when compared to healthy children (Talat et al. 2015). Moreover, taking one or more epileptic drugs makes a significant difference in serum zinc levels, as well as the zinc levels in the serum of patients treated with Phenobarbital, Phenytoin, Carbamazepine, Valproate sodium, Clonazepam, Topiramate, and Primidone, was significantly lower (Saboktakin et al. 2012a).

Table 7.1 Serum levels of zinc and copper based on epileptic drugs usage

	Monotherapy		
	Carbamazepine	Valproate sodium	Polytherapy
Zinc	–	–	–
Copper	+	++	+++

7.2.2.2 Copper and Epilepsy/Seizures

Copper is the body's third most predominant trace element. It has the ability to act on a variety of ion channels, including N-methyl-D-aspartate receptors and voltage-gated calcium channels, both of which play significant roles in neuronal excitability and synaptic communication (Mathie et al. 2006; Saghazadeh et al. 2015). Copper is also involved in catalase and oxidase-type reactions in a number of enzymes (D. K. V. Prasad et al. 2014; Saad et al. 2014). Copper levels have been found to be altered in epilepsy patients, and several antiepileptic medicines have been found to modify these levels (Saghazadeh et al. 2015; Talat et al. 2015). Febrile seizure patients have statistically higher serum copper levels than febrile patients without seizures or meningismus (R. Prasad et al. 2009; Saghazadeh et al. 2015), and in epileptic children compared to healthy ones (Saboktakin et al. 2012b) and in GGE (genetic generalized epilepsy) patients compared to healthy controls (26.69 ± 8.79 μmol/L; 16.64 ± 3.64) ($p < 0.05$) (D. K. V. Prasad et al. 2014). However, when copper is measured in hair, it was found lower in epilepsy patients than in controls (Saghazadeh et al. 2015). Differences are also detected in epileptic patients based on seizure type; among the seizure cases, complex partial seizures (CPS) and generalized seizures (GS) had significantly increased serum copper (Cu) levels than controls (p-0.001) (R. Prasad et al. 2009).

When taking the use of epilepsy drugs into consideration, polytherapy patients have statistically increased copper amounts in their blood than monotherapy patients (Saghazadeh et al. 2015), and the copper level mean in epilepsy long-term treated patients was 180.1±32.4 μg/dl, and in the control group (healthy people) was 114.5±18.5 μg/dl, which was significantly increased in the case group (p<0.001). Furthermore, carbamazepine was found to have more influencing role than sodium valproate in raising serum copper levels in epileptic individuals (Table 7.1) (Talat et al. 2015).In addition, the serum levels of copper were shown to be increased in children with idiopathic epilepsy who were not yet undertaking anti-convulsant therapy (Wojciak et al. 2013). Children with intractable idiopathic epilepsy who were under at least two anti-convulsants were tested for selenium, zinc, and copper levels in a similar study. Serum selenium values in epileptic children were lower than in the control group, although serum copper levels were not significantly different (Seven et al. 2013). Consequently, anti-convulsant therapy may be behind the differences in copper levels (Vitale et al. 2019).Another study assessed some trace element levels in the serum including zinc, copper, iron, and chromium in children before the beginning of pharmacological therapy have demonstrated decreased levels of Zn and Cr in all patients, while copper levels were increased in all patients (Wojciak et al. 2013).

7.2.2.3 Magnesium and Epilepsy/Seizures

Changes in trace element homeostasis, such as magnesium, may contribute to the development of epilepsy. According to experimental research on animals, magnesium (Mg) depletion induces noticeable irritability of the nervous system (Orent et al. 1932; Suter et al. 1958). In animal models of epilepsy, magnesium deficiency lowers seizure thresholds (Toffa et al. 2018; Yuen et al. 2012).And in vitro models of epileptiform activity in neocortical neurons have been developed in magnesium-free media (Yang et al. 2018; Yu et al. 2021).In humans, when comparing epileptic patients to healthy controls, decreased magnesium levels have been found (Abdullahi et al. 2019; Kirkland et al. 2018; Oladipo et al. 2003). The action of the magnesium in the epilepsy still lacks evidence but Mg 2+ is implicated in the processes leading to the pathogenesis of seizures and the neuropathological observed in epilepsy. In some studies, seizure patients had significantly decreased Mg2+ and increased ionized Ca2+/Mg2+ ratios than the matched control group (Mele et al. 2021; Sinert et al. 2007; Wang et al. 2021).

7.2.2.4 Selenium and Epilepsy/Seizures

Selenium is another trace element with effects on the tendency of seizures in human beings and laboratory animals' selenium. In fact, when comparing epileptic patients to healthy controls, selenium levels were significantly lower (Jia et al. 2020; Yüzbaşioğlu et al. 2009). In addition, patients with drug-resistant epilepsy had significantly reduced levels of serum selenium compared to healthy individuals.

7.3 Oligotherapy as a Remedy to Prevent Epileptic Seizures

Like vitamins, trace elements have been the focus of scientific research for several years now. Naturally present in our food, trace elements are also used as an active principle giving rise to oligotherapy whose mechanisms of action, effectiveness, and therapeutic virtues are poorly known and deserve to be defined. Epileptogenesis is a great challenge. Several human and animal research have shown that trace elements, electrolytes, membrane lipid peroxidation, and antioxidant regulation are all important for normal brain activity. Besides, these elements have been linked to the pathophysiology of neuronal excitability, seizure frequency, and antiepileptic medication resistance, either directly or indirectly. Trace minerals and antioxidants may have targeted neuroprotective biological effects when employed in epilepsy patients. Numerous studies have provided evidence for the role of selenium in neuronal susceptibility to excitotoxic injury (Hamed and Abdellah 2004). In neural cell culture, adding selenium in the form of selenite within a physiological range protects against excitotoxins and even reduces primary damage. The neuroprotective impact of selenite is achieved by de novo protein synthesis rather than a direct antioxidant effect

(Savaskan et al. 2003). In addition, studies on the effect of zinc on neurotoxicity have revealed that micromolar zinc concentrations protect cortical neurons against the HO-toxic consequences of glutamate overexposure (Koh and Choi 1988).

A body of preclinical and clinical research indicates that oxidative stress is involved in epilepsy pathogenesis (NazIroğlu et al. 2009; Rowley and Patel 2013). Copper, zinc, and selenium are the most important antioxidant trace elements. For instance, selenium, an essential trace element for humans, has crucial neural antioxidant qualities, as it protects neuronal cell bodies from oxidizing and keeps them in good biological condition (Nazıroğlu and Yürekli 2013). The most essential action of the selenium-dependent enzyme gluthatione peroxidase (GSH-Px) is reducing hydrogen and organic peroxides in the presence of increased GSH (Weber et al. 1991).There is, therefore, a direct link between antioxidant trace elements and the inhibition of the occurrence of epilepsy by oxidative stress. Moreover, the onset and evolution of epilepsy are induced by free radicals (HO-,.O2-,H2O2..) and medications focused on decreasing oxidative stress can improve tissue damage and positively modify the clinical progression of these seizures in an indirect way (Fig. 7.1) (Nazıroğlu and Yürekli 2013).

An animal model in rats is used to indicate that supplementing Se with antiepileptic medicines (AEDs) could be advantageous in treating most of the changes in the brain caused by seizure exposures, and could be recommended as an adjunct therapy with AEDs (Mohammed et al. 2020). The significant difference in the serum levels of certain trace minerals, such as selenium, between healthy dogs and dogs with idiopathic epilepsy may indicate that these nutrients have a role in epilepsy

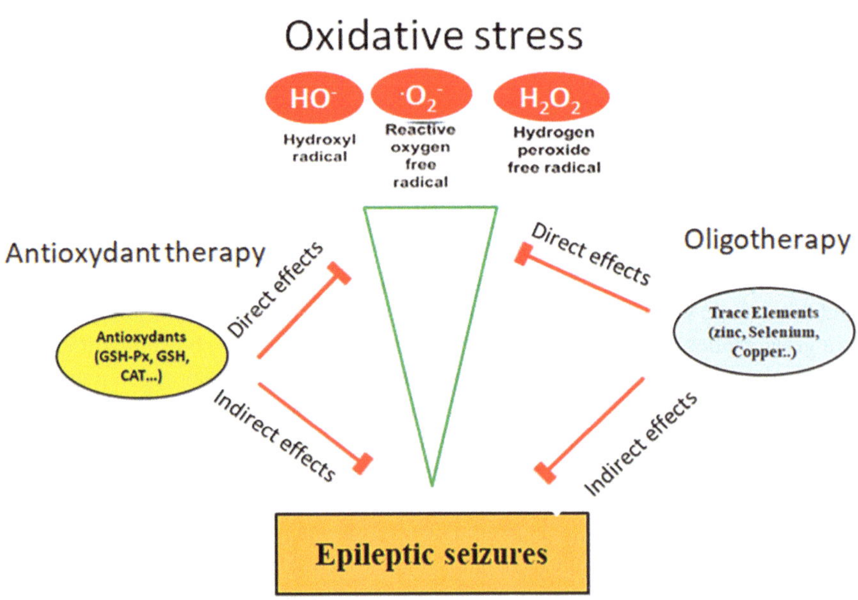

Fig. 7.1 Action of Oligotherapy associated with antioxidants on oxidative stress to limit the occurrence of epileptic seizures

pathogenesis and/or treatment (Vitale et al. 2019). Se supplementation with antiepileptic medications (AED) can enhance most of the effects induced in the brain following seizures (Mohammed et al. 2020) (Mohammed et al. 2020). Recent studies in animal model attempting to use selenium as a treatment showed SeNPs (Selenium nanoparticles) to have the potential to be a promising anticonvulsant medication or therapy regimen (Tawfik et al. 2018; Yuan et al. 2020). Recent data show that the administration of selenium with or without topiramate in humans and animals reduced seizure levels, although antioxidant values were high (Nazıroğlu and Yürekli 2013). The low levels of selenium and GSH-Px were found in epileptic patients (Weber et al. 1991; Yüksel et al. 2001). Selenium-deficient rats were more sensitive to excitotoxicity (Ashrafi et al. 2007). This may mean that supplementation with trace elements, particularly selenium, could restore and counteract the effects of deficiency. Similarly, the administration of selenium to children with refractory epilepsy may help treat children after the discontinuation of anticonvulsant drugs (Oztas et al. 2007; Weber et al. 1991). Thus, the management of epileptic seizures with selenium has a positive effect on the reduction of the blood-brain barrier in male rats (Weber et al. 1991). According to a recent study, higher magnesium consumption in the diet was linked to a lower risk of epilepsy in middle-aged and older males from Eastern Finland (Baek et al. 2018; Yary and Kauhanen 2019). Now, magnesium is prescribed as a supplement. However, large-scale randomized controlled trials are needed to further understand the possible function of magnesium in the treatment of epilepsy. In addition, zinc is a crucial trace element for humans and animals. Numerous studies have shown that zinc regulates specific GABA receptors, and this mechanism is known to help stop seizures (Amiri et al. 2010; André et al. 2010).

7.4 Conclusion

The results of research work presented in this chapter have demonstrated that an alteration of trace elements in the blood and/or in the brain of epileptic patients could be involved in the physiopathology of epilepsy. In this context, oligotherapy can be recommended as a remedy to prevent and reduce the risk of epileptic seizures.

References

Abdullahi I, Watila MM, Shahi N, Nyandaiti Y, Bwala S (2019) Serum magnesium in adult patients with idiopathic and symptomatic epilepsy in Maiduguri, Northeast Nigeria. *Niger J Clin Pract 22*(2):186–193. https://doi.org/10.4103/njcp.njcp_252_18

Amiri M, Farzin L, Moassesi ME, Sajadi F (2010) Serum trace element levels in febrile convulsion. *Biol Trace Elem Res 135*(1–3):38–44. https://doi.org/10.1007/S12011-009-8487-6

André VM, Cepeda C, Vinters HV, Huynh M, Mathern GW, Levine MS (2010) Interneurons, GABA currents, and subunit composition of the GABA A receptor in type I and type II cortical dysplasia. *Epilepsia 51*(SUPPL. 3):166–170. https://doi.org/10.1111/j.1528-1167.2010.02634.x
Arul J, Kommu PPK, Kasinathan A, Ray L, Krishnan L (2020) Zinc status and febrile seizures: results from a cross-sectional study. *J Neurosci Rural Pract 11*(4):597–600. https://doi.org/10.1055/S-0040-1715992
Ashrafi MR, Shams S, Nouri M, Mohseni M, Shabanian R, Yekaninejad MS, Chegini N, Khodadad A, Safaralizadeh R (2007) A probable causative factor for an old problem: selenium and glutathione peroxidase appear to play important roles in epilepsy pathogenesis. *Epilepsia 48*(9):1750–1755. https://doi.org/10.1111/J.1528-1167.2007.01143.X
Baek SJ, Byeon JH, Eun SH, Eun BL, Kim GH (2018) Risk of low serum levels of ionized magnesium in children with febrile seizure. *BMC Pediatr 18*:1. https://doi.org/10.1186/S12887-018-1271-Z
Carl GF, Critchfield JW, Thompson JL, McGinnis LS, Wheeler GA, Gallagher BB, Holmes GL, Hurley LS, Keen CL (1989) Effect of kainate-induced seizures on tissue trace element concentrations in the rat. *Neuroscience 33*(1):223–227. https://doi.org/10.1016/0306-4522(89)90323-0
Chwiej J, Winiarski W, Ciarach M, Janeczko K, Lankosz M, Rickers K, Setkowicz Z (2008) The role of trace elements in the pathogenesis and progress of pilocarpine-induced epileptic seizures. *J Biol Inorganic Chem 13*(8):1267–1274. https://doi.org/10.1007/S00775-008-0411-6
Doboszewska U, Młyniec K, Wlaź A, Poleszak E, Nowak G, Wlaź P (2019) Zinc signaling and epilepsy. *Pharmacol Ther* 193:156–177. https://doi.org/10.1016/j.pharmthera.2018.08.013
Doretto MC, Simões S, Paiva AMR, Osório-Neto E (2002) Zinc, magnesium and copper profiles in three experimental models of epilepsy. *Brain Res 956*(1):166–172. https://doi.org/10.1016/S0006-8993(02)03517-5
Eissa MA, Abdulghani KO, Nada MA, Elkhawas HM, Shouman AE, Ahmed NS (2020) Serum zinc and copper levels in a sample of Egyptian epileptic children. Egypt J Neurol Psychiatry Neurosurg 56:1–9. https://doi.org/10.1186/s41983-020-00210-2
El Khiat A, Kissani N, Abbaoui A, Chraa M, Loukid M, Gamrani H, Hilali MK (2021) Behaviour and therapeutic itinerary of epileptic patients in Marrakech city and surrounding regions in South Morocco. *Acta Epileptologica 3*(1):1–6. https://doi.org/10.1186/S42494-021-00061-1/TABLES/4
Farahani HN, Ashthiani AR, Masihi MS (2013) Study on serum zinc and selenium levels in epileptic patients. *Neurosci J 18*:2
Fukahori M, Itoh M, Oomagari K, Kawasaki H (1988) Zinc content in discrete hippocampal and amygdaloid areas of the epilepsy (El) mouse and normal mice. *Brain Res 455*(2):381–384. https://doi.org/10.1016/0006-8993(88)90099-6
Ganesh R, Janakiraman L (2008) Serum zinc levels in children with simple febrile seizure. *Clin Pediatr 47*(2):164–166. https://doi.org/10.1177/0009922807306165
Hamed SA, Abdellah MM (2004) Trace elements and electrolytes homeostasis and their relation to antioxidant enzyme activity in brain hyperexcitability of epileptic patients. *J Pharmacol Sci 96*(4):349–359. https://doi.org/10.1254/JPHS.CRJ04004X
Hirate M, Takeda A, Tamano H, Enomoto S, Oku N (2002) Distribution of trace elements in the brain of EL (epilepsy) mice. *Epilepsy Res 51*(1–2):109–116. https://doi.org/10.1016/S0920-1211(02)00098-0
Jia W, Song Y, Yang L, Kong J, Boczek T, He Z, Wang Y, Zhang X, Hu H, Shao D, Tang H, Xia L, Xu X, Guo F (2020) The changes of serum zinc, copper, and selenium levels in epileptic patients: a systematic review and meta-analysis. *Expert Rev Clin Pharmacol 13*(9):1047–1058. https://doi.org/10.1080/17512433.2020.1816821
Khan MZ (2016) A possible significant role of zinc and GPR39 zinc sensing receptor in Alzheimer disease and epilepsy. *Biomed Pharmacother = Biomedecine & Pharmacotherapie 79*:263–272. https://doi.org/10.1016/J.BIOPHA.2016.02.026
Kheradmand Z, Yarali B, Zare A, Pourpak Z, Shams S, Ashrafi MR (2014) Comparison of serum zinc and copper levels in children and adolescents with intractable and controlled epilepsy. *Iran J Child Neurol 8*(3):49. /pmc/articles/PMC4135281/

Kirkland AE, Sarlo GL, Holton KF (2018) The role of magnesium in neurological disorders. *Nutrients 10*:6. https://doi.org/10.3390/NU10060730

Kissani N, Nafia S, El Khiat A, Bengamara N, Maiga Y, Sogoba Y, Ibrahim AA, Agbetou ME, Massi Daniel G, Assogba K, Matar Gaye N, Kuate Tegueu C, Hussein Ragab A, Razafimahefa J, Wilmshurst J, Naidoo A, Jabang JN, Watila M (2021) Epilepsy surgery in Africa: state of the art and challenges. *Epilepsy Behav 118*:107910. https://doi.org/10.1016/J.YEBEH.2021.107910

Koh J, Choi DW (1988) Zinc alters excitatory amino acid neurotoxicity on cortical neurons. *J Neurosci 8*(6):2164–2171. https://doi.org/10.1523/jneurosci.08-06-02164.1988

Lee JH, Kim JH (2012) Comparison of serum zinc levels measured by inductively coupled plasma mass spectrometry in preschool children with febrile and afebrile seizures. *Ann Lab Med 32*(3):190–193. https://doi.org/10.3343/ALM.2012.32.3.190

Lynes MA, Kang YJ, Sensi SL, Perdrizet GA, Hightower LE (2007) Heavy metal ions in normal physiology, toxic stress, and cytoprotection. *Ann N Y Acad Sci 1113*:159–172. https://doi.org/10.1196/annals.1391.010

Mathie A, Sutton GL, Clarke CE, Veale EL (2006) Zinc and copper: pharmacological probes and endogenous modulators of neuronal excitability. *Pharmacol Ther 111*(3):567–583. https://doi.org/10.1016/J.PHARMTHERA.2005.11.004

Mehri A (2020) Trace elements in human nutrition (II)—an update. *Int J Prevent Med*. https://doi.org/10.4103/ijpvm.IJPVM_48_19

Mele M, Vieira R, Correia B, De Luca P, Duarte FV, Pinheiro PS, Duarte CB (2021) Transient incubation of cultured hippocampal neurons in the absence of magnesium induces rhythmic and synchronized epileptiform-like activity. *Sci Rep* 11(1):1–14. https://doi.org/10.1038/s41598-021-90486-y

Mohammed HS, Aboul Ezz HS, Zedan A, Ali MA (2020) Electrophysiological and neurochemical assessment of selenium alone or combined with carbamazepine in an animal model of epilepsy. *Biol Trace Element Res 195*(2):579–590. https://doi.org/10.1007/s12011-019-01872-7

Namakin K, Zardast M, Sharifzadeh G, Bidar T, Zargarian S (2016) SerumTrace elements in febrile seizure: a case-control study. *Iran J Child Neurol 10*(3):57. /pmc/articles/PMC4928617/

NazIroğlu M, Kutluhan S, Uğuz AC, Çelik Ö, Bal R, Butterworth PJ (2009) Topiramate and vitamin e modulate the electroencephalographic records, brain microsomal and blood antioxidant redox system in pentylentetrazol-induced seizure of rats. *J Membr Biol 229*(3):131–140. https://doi.org/10.1007/s00232-009-9177-1

Nazıroğlu M, Yürekli VA (2013) Effects of antiepileptic drugs on antioxidant and oxidant molecular pathways: focus on trace elements. *Cell Mol Neurobiol 33*(5):589–599. https://doi.org/10.1007/S10571-013-9936-5

Oladipo OO, Ajala MO, Okubadejo N, Danesi MA, Afonja OA (2003) Plasma magnesium in adult Nigerian patients with epilepsy. *Niger Postgrad Med J 10*(4):234–237. https://europepmc.org/article/med/15045017

Orent ER, Kruse HD, McCollum EV (1932) Studies on magnesium deficiency in animals. *Am J Physiol-LegacyContent101*(3):454–461.https://doi.org/10.1152/AJPLEGACY.1932.101.3.454

Oztas B, Akgul S, research, F. S.-B. trace element, & 2007, undefined (2007) Gender difference in the influence of antioxidants on the blood–brain barrier permeability during pentylenetetrazol-induced seizures in hyperthermic rat pups. *Springer 118*(1):77–83. https://doi.org/10.1007/s12011-007-0020-1

Prakash A, Bharti K, Majeed ABA (2015) Zinc: indications in brain disorders. *Fundamental Clin Pharmacol 29*(2):131–149. https://doi.org/10.1111/FCP.12110

Prasad DKV, Shaheen U, Satyanarayana U, Surya Prabha T, Jyothy A, Munshi A (2014) Association of serum trace elements and minerals with genetic generalized epilepsy and idiopathic intractable epilepsy. *Neurochem Res 39*(12):2370–2376. https://doi.org/10.1007/S11064-014-1439-3

Prasad R, Singh A, Das BK, Upadhyay RS, Singh TB, Mishra OP (2009) Cerebrospinal fluid and serum zinc, copper, magnesium and calcium levels in children with idiopathic seizure. *J Clin Diagn Res 3*(6):1841–1846. http://www.jcdr.net/back_issues.asp?issn=0973-709x&year=2009&month=December&volume=3&issue=6&page=1841-146&id=457

Rowley S, Patel M (2013) Mitochondrial involvement and oxidative stress in temporal lobe epilepsy. *Free Radic Biol Med 62*:121–131. https://doi.org/10.1016/J.FREERADBIOMED.2013.02.002

Saad K, Hammad E, Hassan AF, Badry R (2014) Trace element, oxidant, and antioxidant enzyme values in blood of children with refractory epilepsy. *Int J Neurosci 124*(3):181–186. https://doi.org/10.3109/00207454.2013.831851

Saboktakin L, Barzegar M, Ghorbani A, Jo H, Emamalizadeh M (2012a) Study on serum Copper and Zinc level of children with epilepsy during long term therapy with anticonvulsants. *Life Sci J 9*(4):1097–8135. http://www.lifesciencesite.comhttp//www.lifesciencesite.com.188

Saboktakin L, Barzegar M, Hagh Jo AG, Emamalizadeh M (2012b) Study on serum copper and zinc level of children with epilepsy during long term therapy with anticonvulsants. *Life Sci J 9*(4):1250–1254. http://citeseerx.ist.psu.edu/viewdoc/download?doi=10.1.1.381.4644&rep=rep1&type=pdf

Saghazadeh A, Mahmoudi M, Meysamie A, Gharedaghi M, Zamponi GW, Rezaei N (2015) Possible role of trace elements in epilepsy and febrile seizures: a meta-analysis. *Nutr Rev 73*(11):760–779. https://doi.org/10.1093/nutrit/nuv026

Saiki M, Leite REP, Genezini FA, Grinberg LT, Ferretti REL, Farfel JM, Suemoto C, & Pasqualucci CA (2012) *Trace element concentration differences in regions of human brain by INAA*. https://doi.org/10.1007/s10967-012-2055-9

Salah ON, Abdelraouf ER, Abdelhameed MH, Dawood AA, Hashish AF, Kilany A, Helal SI (2014) Assessment of the level of GABA and some trace elements in blood in children who suffer from familial febrile convulsions. *Open Access Maced J Med Sci 2*(1):68–73. https://doi.org/10.3889/oamjms.2014.012

Salehiomran MR, Mahzari M (2013) Zinc status in febrile seizure: a case-control study. *Iran J Child Neurol 7*(4):20–23. https://pubmed.ncbi.nlm.nih.gov/24665313/

Savaskan NE, Bräuer AU, Kühbacher M, Eyüpoglu IY, Kyriakopoulos A, Ninnemann O, Behne D, Nitsch R (2003) Selenium deficiency increases susceptibility to glutamate-induced excitotoxicity. *FASEB J 17*(1):112–114. https://doi.org/10.1096/FJ.02-0067FJE

Scott RA, Lhatoo SD, Sander JWAS (2001) Le traitement de l'épilepsie dans les pays en développement : quelles pistes pour demain ? *Bulletin of the World Health Organization 79*(4):344–351

Seven M, Basaran SY, Cengiz M, Unal S, Yuksel A (2013) Deficiency of selenium and zinc as a causative factor for idiopathic intractable epilepsy. *Epilepsy Res 104*(1–2):35–39. https://doi.org/10.1016/J.EPLEPSYRES.2012.09.013

Sinert R, Zehtabchi S, Desai S, Peacock P, Altura BT, Altura BM (2007) Serum ionized magnesium and calcium levels in adult patients with seizures. *Scand J Clin Lab Invest 67*(3):317–326. https://doi.org/10.1080/00365510601051441

Suter C, Klingman WO, Boggs D, Marks RD, Coplinger CB, Randolph V (1958) Sound-induced seizures in animals: magnesium deficiency and sound-induced seizures in rats. *Neurology 8*:125–128. https://doi.org/10.1212/WNL.8.Suppl_1.125

Takeda A (2004) Analysis of brain function and prevention of brain diseases: the action of trace metals. *J Health Sci 50*(5):429–442. https://doi.org/10.1248/jhs.50.429

Takeda A, Hanajima T, Ijiro H, Ishige A, Iizuka S, Okada S, Oku N (1999) Release of zinc from the brain of El (epilepsy) mice during seizure induction. *Brain Res 828*(1–2):174–178. https://doi.org/10.1016/S0006-8993(99)01358-X

Talat MA, Ahmed A, Mohammed L (2015) Serum levels of zinc and copper in epileptic children during long-term therapy with anticonvulsants. *Neurosci J 20*(4):341–345. https://doi.org/10.17712/NSJ.2015.4.20150336

Tawfik KM, Moustafa YM, El-Azab MF (2018) Neuroprotective mechanisms of sildenafil and selenium in PTZ-kindling model: Implications in epilepsy. *Eur J Pharmacol 833*:131–144. https://doi.org/10.1016/j.ejphar.2018.05.035

Toffa DH, Kpadonou C, Gams Massi D, Ouedraogo M, Sow AD, Samb MNEA (2018) Le magnésium et le calcium réduisent la sévérité des troubles de la mémoire spatiale pour le modèle kaïnique d'épilepsie mésiale temporale chez la souris. *Canadian J Physiol Pharmacol 96*(11):1132–1144. https://doi.org/10.1139/CJPP-2018-0020

Vitale S, Hague DW, Foss K, de Godoy MC, Selmic LE (2019) Comparison of serum trace nutrient concentrations in epileptics compared to healthy dogs. *Front Vet Sci 6*. https://doi.org/10.3389/fvets.2019.00467

Wandt VK, Winkelbeiner N, Bornhorst J, Witt B, Raschke S, Simon L, Ebert F, Kipp AP, Schwerdtle T (2021) Redox biology review article A matter of concern—Trace element dyshomeostasis and genomic stability in neurons. *Redox Biol 41*(January):101877. https://doi.org/10.1016/j.redox.2021.101877

Wang C, Li Y, Li Y, Du L, Zhang J, Li N, Hu X, Zhang W, Xie N, Ming L (2021) FAM134B-Mediated ER-Phagy in Mg2+-Free solution-induced mitochondrial calcium homeostasis and cell death in epileptic hippocampal neurons. *Neurochem Res 46*(9):2485–2494. https://doi.org/10.1007/S11064-021-03389-9

Weber GF, Maertens P, Meng X, Pippenger CE (1991) Glutathione peroxidase deficiency and childhood seizures. *The Lancet 337*(8755):1443–1444. https://doi.org/10.1016/0140-6736(91)93130-2

Wojciak RW, Mojs E, Stanislawska-Kubiak M, Samborski W (2013) The serum zinc, copper, iron, and chromium concentrations in epileptic children. *Epilepsy Res 104*(1–2):40–44. https://doi.org/10.1016/J.EPLEPSYRES.2012.09.009

Yang Q, Zheng F, Hu Y, Yang Y, Li Y, Chen G, Wang W, He M, Zhou R, Ma Y, Xu D, Tian X, Gao X, Wang Q, Wang X (2018) ZDHHC8 critically regulates seizure susceptibility in epilepsy. *Cell Death Dis 9*:8. https://doi.org/10.1038/s41419-018-0842-0

Yary T, Kauhanen J (2019) Dietary intake of magnesium and the risk of epilepsy in middle-aged and older Finnish men: A 22-year follow-up study in a general population. *Nutrition 58*:36–39. https://doi.org/10.1016/j.nut.2018.06.019

Yu Y, Du L, Zhang J (2021) Febrile seizure-related miR-148a-3p exerts neuroprotection by promoting the proliferation of hippocampal neurons in children with temporal lobe epilepsy. *Dev Neurosci 43*(5):312–320. https://doi.org/10.1159/000518352

Yuan X, Fu Z, Ji P, Guo L, Al-Ghamdy AO, Alkandiri A, Habotta OA, Moneim AEA, Kassab RB (2020) Selenium nanoparticles pre-treatment reverse behavioral, oxidative damage, neuronal loss and neurochemical alterations in pentylenetetrazole-induced epileptic seizures in mice. *Int J Nanomed 15*:6339–6353. https://doi.org/10.2147/IJN.S259134

Yuen A, research, J. S.-E, 2012, undefined (2012) Can magnesium supplementation reduce seizures in people with epilepsy? A hypothesis. *Elsevier 100*(1–2):152–156. https://doi.org/10.1016/j.eplepsyres.2012.02.004

Yüksel A, Cengiz M, Seven M, Ulutin T (2001) Changes in the antioxidant system in epileptic children receiving antiepileptic drugs: two-year prospective studies. *J Child Neurol 16*(8):603–606. https://doi.org/10.1177/088307380101600814

Yüzbaşioğlu A, Karataş H, Gürsoy-Özdemir Y, Saygi S, Akalan N, Söylemezoğlu F, Dalkara T, Kocaefe YÇ, Özgüç M (2009) Changes in the expression of selenoproteins in mesial temporal lobe epilepsy patients. *Cell Mol Neurobiol 29*(8):1223–1231. https://doi.org/10.1007/S10571-009-9418-Y

Chapter 8
Blood Markers of Oxidative Stress in Patients with Amyotrophic Lateral Sclerosis

Sarah Hassan, Mario Eid, Ahmad Hassan, and Samer El Hayek

Abstract Amyotrophic lateral sclerosis (ALS) is the most common motor neuron disease in adult populations and is characterized by progressive cortical and spinal motor neuron degeneration. Several processes are implicated in the pathogenesis of the disease, including neuronal excitotoxicity, neuroinflammation, abnormal protein aggregation, and oxidative stress. This chapter reviews the role that oxidative stress plays in the development and progression of ALS, research into oxidative stress biomarkers that can potentially aid in diagnosis and monitoring disease progression, and therapeutics that aim to target oxidative stress mechanisms.

Keywords Amyotrophic lateral sclerosis · ALS · Oxidative stress · Biomarkers

8.1 Introduction

Amyotrophic lateral sclerosis (ALS) is one of the most common neurodegenerative diseases, with a prevalence between 4.1 and 8.4 per 100,000 persons and a median age of onset between 51 and 66 years (Longinetti and Fang 2019). ALS is characterized by a selective and progressive deterioration of upper and lower motor neurons (Hulisz 2018). This leads to progressive weakness and atrophy of skeletal muscles, typically resulting in quadriplegia and fatal respiratory failure during the later stages of the disease. ALS is a rapidly progressive disease, and a majority of patients die from respiratory paralysis within 3 to 5 years after the onset of symptoms (Sun et al.

S. Hassan
The Chicago School of Professional Psychology, Chicago, IL, USA

M. Eid
Faculty of Medicine, Lebanese University, Hadath, Lebanon

A. Hassan
Department of Psychiatry, Washington University School of Medicine, St. Louis, MO, USA

S. El Hayek (✉)
Medical Department, Erada Center for Treatment and Rehab, Dubai, United Arab Emirates

W. Mohamed, R. Sandhir (eds.), *Trace Elements in Brain Health and Diseases*, Nutritional Neurosciences, https://doi.org/10.1007/978-981-99-1513-2_8

2020). Patients with ALS commonly endorse symptoms such as fatigue, weakness, pain, shortness of breath, sleep disturbances, speech problems, cognitive changes, as well as depression and other mood changes. All these symptoms can have a substantially debilitating effect on daily life (Brizzi et al. 2020). Patients and caregivers alike often experience devastating financial impacts associated with treatment cost, lack of insurance coverage, and travel. In about 10% of patients with ALS, family history suggests an autosomal dominant pattern of inheritance, and these patients are classified as having familial ALS. The remaining 90% of patients have no affected relatives and are classified as patients with sporadic ALS (Masrori and Van Damme 2020).

Several possible pathogenic mechanisms contributing to the destruction of motor neurons have been identified in ALS. These include oxidative stress, mitochondrial dysfunction, protein aggregation, inflammation, apoptosis, free radical generation, and excitotoxicity of neurons (Hinchcliffe and Smith 2017). Oxidative stress describes the condition in which pro- and antioxidant balance in cells is disturbed, leading to the production of cell-damaging reactive oxygen (ROS) and nitrogen (RNS) species.

The exact cause of ALS remains undetermined, and it is not yet clear whether oxidative stress plays a causative role in the development of the disease, or if it is a downstream effect of other pathological processes seen in ALS, such as protein aggregation. Research into oxidative stress mechanisms has provided further insight into the pathogenesis of the disease. In 1993, Rosen and colleagues identified a link between ALS and mutations in the superoxide dismutase 1 gene (*SOD1*) (Rosen et al. 1993). The *SOD1* gene encodes the essential antioxidant enzyme Cu, Zn superoxide dismutase (SOD-1), which converts superoxide radicals into peroxide (Berdyński et al. 2022). In patients with ALS, mutations in the *SOD1* gene are found in the exon region, supporting the hypothesis that oxidative stress is a driving mechanism behind the loss of motor neurons in ALS (Bond et al. 2018). Furthermore, in recent years, increasing evidence has identified redox dysregulation as an important contributor to disease pathogenesis (Jagaraj et al. 2021).

8.2 Role of Oxidative Stress in Amyotrophic Lateral Sclerosis

Pathophysiological processes commonly observed in neurodegenerative diseases include oxidative stress, protein aggregation, and impaired DNA repair mechanisms. In ALS, each of these processes was found to play an important role in the pathogenesis of the disease. Neurons are highly vulnerable to oxidative stress, and antioxidant mechanisms typically decrease in effectiveness with age (Tarafdar and Pula 2018). It has long been known that the biomarkers of oxidative stress are elevated in neurodegenerative diseases (Singh et al. 2019). Recently, several mechanisms of neuronal oxidative stress damage have been identified in patients with ALS

(Harley et al. 2021). Alternatively, abnormal protein folding and aggregation is another pathological hallmark of neurodegenerative diseases: α-synuclein in Parkinson's disease, β-amyloid in Alzheimer's disease, and transactive response DNA-binding protein-43 (TDP-43) in ALS and frontotemporal dementia (Hillen 2019; Fields et al. 2019; Suk and Rousseaux 2020). Researchers believe that oxidative stress is both causative and consecutive to protein aggregation with an abundance of literature supporting this theory (Lévy et al provide a detailed literature review on the cyclical interaction between oxidative stress and protein aggregation (Lévy et al. 2019)).

In ALS, increased oxidative stress (due to aging or mitochondrial dysfunction) induces the release of cytochrome C from neuronal mitochondria, which promotes the aberrant oligomerization of proteins such as SOD-1 and TDP-43, ultimately resulting in neurodegeneration (Lévy et al. 2019). This is the major pathological hallmark of both sporadic and familial ALS (Hosaka et al. 2021). These protein aggregates have been identified in animal models, as well as postmortem spinal cord samples of patients with ALS (Benkler et al. 2018). Misfolded SOD-1 obtained from cerebrospinal fluid (CSF) samples of patients with ALS showed significant toxicity towards motor neuron-like cells, and this toxicity was ameliorated by the removal of the misfolded proteins via immunoprecipitation (Tokuda et al. 2019). Although protein aggregates can be found throughout the nervous system of patients with ALS, spreading in a prion-like fashion, neuronal degeneration seems to be specific to motor neurons, and this selectivity is still not fully understood (Ayers and Cashman 2018).

TDP-43 is an RNA/DNA binding protein, which plays a role in RNA metabolism and forms hyper-phosphorylated and ubiquitinated cytoplasmic inclusions in the motor neurons of patients with ALS (de Boer et al. 2021). TDP-43 neuronal inclusions are considered a pathological hallmark of ALS and are found in nearly all patients, regardless of the ALS type and TDP-43 mutational status, indicating the pivotal role TDP-43 plays in disease pathogenesis (Prasad et al. 2019). Biochemical studies have demonstrated that prolonged oxidative stress induces the aggregation of TDP-43, via the promotion of protein acetylation (Cohen et al. 2015; Goh et al. 2018). However, it is thought that such aggregation occurs at a later stage in disease development, and targeting protein aggregates may not lead to a substantial therapeutic impact (Suk and Rousseaux 2020).

Oxidative damage to DNA has also been implicated in the pathogenesis of ALS. An important biomarker and product of oxidative stress is 8-hydroxy-2′-deoxyguanosine (8-OHdG); its levels are shown to increase with aging and other diseases in which oxidative stress has been implicated (Qing et al. 2019). The elevated levels of 8-OHdG have been observed in motor neurons of G93A-mSOD1 mutant mice, an animal model of familial ALS (Hosaka et al. 2021). A study on G93A-mSOD1 transgenic mice found that motor neuron degeneration is characterized by swelling of the soma and mitochondria, the formation of DNA single-strand breaks, p53 activation, and subsequent cellular apoptosis (Martin et al. 2007). 8-OHdG levels were also found to be increased in postmortem spinal cord tissue extracts of patients with ALS (Kim et al. 2020).

Dysfunctions in DNA repair mechanisms are also hypothesized to play an important role in ALS pathogenesis. Early studies performed on skin fibroblast cells obtained from patients with ALS found that the cells had a decreased ability to repair alkylation lesions in DNA caused by oxidative stress, specifically at apurinic and apyrimidinic sites (Tandan et al. 1987; Bradley et al. 1987). This led researchers to investigate the role of apurinic apyrimidinic endonuclease 1 (APE1)/redox effector factor 1 (Ref-1) in ALS. APE1/Ref-1 is a crucial enzyme involved in the base excision repair (BER) pathway, which plays a role in repairing damage caused to DNA by oxidative stress, and is important in preserving genome stability. It is also a redox-dependent regulator of transcription factors such as activator protein-1 (AP-1), nuclear factor-κB (NF-κB), signal transducer and activator of transcription 3 (STAT3), and hypoxia-inducible factor 1 alpha (HIF-1α), serving a vital role in cell signaling, senescence, and inflammation pathways (Oliveira et al. 2022). One study noted that patients with sporadic ALS had reduced frontal cortical levels and activity of APE1 (Kisby et al. 1997). However, another study found elevated levels of APE1/Ref-1 in the brain and spinal cord samples of patients with ALS (Shaikh and Martin 2002). This led researchers to hypothesize that damage to DNA may be an upstream mechanism for the degeneration of motor neurons in ALS, and that the BER pathway is activated in response to DNA damage. As the disease progresses, however, the prolonged stimulation of the BER pathway and other DNA repair mechanisms leads to exhaustion and depletion (Thakur et al. 2014). These findings support the hypothesis that impairment of DNA repair mechanisms and subsequent oxidative damage of DNA are driving mechanisms of motor neuron deterioration in ALS.

8.3 Oxidative Stress Biomarkers in Amyotrophic Lateral Sclerosis

Similar to most neurodegenerative diseases, ALS is diagnosed based on clinical symptoms (Masrori and Van Damme 2020). Typically, both upper and lower motor neuron signs can be seen on neurological exams. A summary of common exam findings is provided in Table 8.1. A definitive diagnosis of a neurodegenerative disease requires a histological examination of the spinal cord and brain tissue, which is

Table 8.1 Common clinical findings seen in patients with ALS

Upper motor neuron findings	Lower motor neuron findings	Other common findings
Positive Babinski sign Hyperreflexia Muscle spasticity	Muscle weakness Muscle atrophy Fasciculations Foot drop	Pain Dysarthria Dysphagia Depression Emotional lability Sleep disturbances Constipation

impossible except postmortem (Hosaka et al. 2021). Consequently, researchers have looked into biomarkers - biological molecules in the blood, tissues, or body fluids that serve as indicators of normal or pathogenic processes - to assist in the screening, diagnosing, and treating of neurodegenerative disorders (Hansson 2021). While the exact cause of ALS remains unclear, it is now well-supported that downstream pathophysiological events of the disease include oxidative stress, neuroinflammation, protein aggregation, and glutamate-mediated excitotoxicity (Limanaqi et al. 2020). Research into ALS biomarkers may lead to new strategies to diagnose the disease, determine its extent and progression, and identify novel therapeutic targets. This section introduces several potential biomarkers which have been investigated in the context of ALS.

Several promising biomarker candidates for ALS have emerged from recent research. For instance, a recent study found elevated levels of serum 8-OHdG in affected patients (Blasco et al. 2017). The elevated levels of 8-OHdG have also been found in the CSF, plasma, and urine of patients with both sporadic and familial ALS (Hosaka et al. 2021). Urine 8-OHdG levels were inversely correlated to the rate of change of the revised ALS functional rating scale (ALSFRS-R) score, which is used to measure and track physical function over time. Urine 8-OHdG levels were also inversely correlated to the forced vital capacity of the lungs, indicating that it may be a useful biomarker for not only diagnosis but also monitoring disease progression (Bogdanov et al. 2000).

4-hydroxynonenal (HNE) and malondialdehyde (MDA) are the products of lipid peroxidation, one of the many results of oxidative stress, and have been investigated and utilized as biomarkers for oxidative stress in various diseases (Singh et al. 2014). Studies found elevated serum levels of MDA in patients with ALS. The serum levels of HNE have been shown to correlate with the extent of disease in ALS patients (Hosaka et al. 2021; Blasco et al. 2017; Baillet et al. 2010).

3-Nitrotyrosine (3-NT) is another promising oxidative stress biomarker that is formed as a result of protein-bound and free-tyrosine nitration by reactive peroxynitrite molecules. It has been investigated as a biomarker in various neurodegenerative diseases (Bandookwala and Sengupta 2020). Patients with both familial and sporadic ALS were found to have elevated levels of 3-NT in their spinal cord (Beal et al. 1997).

The levels and activity of SOD-1 can potentially be used as reliable biomarkers in ALS. Early studies found elevated activity of SOD-1 in the plasma and red blood cells of patients with ALS (Ihara et al. 1995; Moumen et al. 1997). One recent study identified misfolded SOD-1 proteins in the CSF of patients with sporadic ALS (Tokuda et al. 2019). The same study demonstrated that these misfolded proteins pose considerable toxicity toward motor neuron-like hybrid cells.

Other biomarkers that have been studied in ALS include serum total antioxidant status (TAS) – which provides information about all antioxidants in the blood serum – and erythrocyte glutathione peroxidase (GPX) - an antioxidant enzyme responsible for reducing hydrogen peroxide and peroxide radicals. The previously mentioned study by Blasco and colleagues also identified increased serum levels of MDA in ALS patients, and a significantly decreased TAS (Blasco et al. 2017).

Alternatively, GPX did not, so far, serve as a reliable oxidative stress biomarker, as previous studies showed inconsistent results, with either reduced activity or no change in activity in patients with ALS (Hosaka et al. 2021).

Lastly, 15-F_{2t}-isoprostane (IsoP), produced via a free radical-catalyzed mechanism from arachidonic acid, has been extensively studied as a useful oxidative stress biomarker in various diseases (Milne et al. 2015). One study found elevated urinary levels of IsoP in patients with sporadic ALS (Mitsumoto et al., 2008).

While these studies have shed light on the importance of oxidative stress and inflammatory mechanisms in ALS pathogenesis, they were often limited by inconsistent results or small sample sizes. A summary of potential ALS biomarkers and their biological function is provided in Table 8.2. Further research is needed to identify biomarkers that may eventually be used to assist in the diagnosis and monitoring of disease progression.

8.4 Treatment Interventions Targeting Oxidative Stress

Many experimental drugs were found to delay ALS progression in preclinical studies but failed to demonstrate efficacy in human clinical trials (Petrov et al. 2017). This section introduces several key therapeutics that have been investigated in the

Table 8.2 Potential biomarkers in ALS

Biomarker	Physiological function	Setting	References
8-OHdG	A product of oxidative damage to 2′-deoxyguanosine	CSF, plasma, urine	Bogdanov et al. (2000), Devos et al. (2019), Hosaka et al. (2021)
MDA	Highly toxic aldehyde molecule, which is a product of polyunsaturated fatty acid peroxidation	Plasma	Baillet et al. (2010), Blasco et al. (2017), Hosaka et al. (2021), Miana-Mena et al. (2011), Singh et al. (2014)
HNE	Highly toxic end-product of lipid peroxidation, implicated in tissue damage and plays an important role in the pathogenesis of several diseases	CSF, plasma	Baillet et al. (2010), Blasco et al. (2017), Devos et al. (2019), Hosaka et al. (2021), Singh et al. (2014)
3-NT	Product of protein-bound and free tyrosine nitration by reactive peroxynitrite molecules	CSF	Bandookwala and Sengupta (2020), Beal et al. (1997), Tohgi et al. (1999)
SOD-1	An antioxidant enzyme found in the cytosol that catalyzes the transformation of the superoxide anion into hydrogen peroxide	CSF, plasma, red blood cells	Ihara et al. (1995), Moumen et al. (1997), Tokuda et al. (2019)
GPX	An antioxidant enzyme that enzymatically reduces hydrogen peroxide to water	Serum	Blasco et al. (2017), Hosaka et al. (2021)
IsoP	A prostaglandin F2-like compound generated by the free radical-catalyzed peroxidation of arachidonic acid	Urine	Devos et al. (2019), Milne et al. (2015), Mitsumoto et al. (2008)

context of ALS. The first FDA-approved therapy for ALS did not target oxidative stress; rather, its mechanism targeted excitotoxicity seen in the disease pathogenesis. In 1995, riluzole, a glutamatergic neurotransmission inhibitor, was approved by the FDA for the treatment of ALS (Gaber et al. 2016). By acting as an antagonist for glutamate, riluzole is thought to reduce neuronal excitability, which has been proposed to partially mediate motor neuron death, by blocking persistent sodium currents, leading to increased life expectancy (King et al. 2016). However, the effect of riluzole is limited as it extends the lifespan of patients with ALS by about three months and has no significant effect on muscular function (Nightingale 2020).

Riluzole remained the only drug therapy for ALS approved by the FDA until May 2017, when the FDA approved edaravone, a strong antioxidant that prevents neuronal death by oxidative stress mechanisms (Bhandari et al. 2018). An in vitro biochemical study showed that edaravone functions as a peroxynitrite scavenger, and a powerful oxidant and nitrating agent of lipid peroxyl radicals (Fujisawa and Yamamoto 2016). An open-label phase II study in patients with ALS found that edaravone decreased the concentration of the oxidative stress biomarker 3-NT and resulted in a significant decline in the ALSFRS-R score six months following the administration of the drug (Okada et al. 2018). Before that, the concentration of 3-NT in the CSF was also found to be reduced to near-undetectable levels at the end of a six-month treatment period with edaravone (Yoshino and Kimura 2006). More recently, a randomized-controlled trial conducted on 45 patients with ALS, 22 of whom received edaravone, found a statistically significant increase in the median survival duration in the edaravone group (49 months) compared to the placebo group (25 months) (Houzen et al., 2021). Currently, edaravone is the most effective medication available for ALS, although its clinical efficacy remains limited.

Widely-known antioxidants, such as vitamins C and E, were investigated to determine their effectiveness at treating or delaying the onset of ALS. Animal studies showed that, although supplementation with vitamin C failed to daily the onset of ALS, it significantly slowed the progression of muscle paralysis (Padayatty et al. 2003). However, several case-control studies failed to demonstrate a relationship between vitamin C intake and ALS risk or progression (Fitzgerald et al. 2013). Alternatively, a pooled prospective study analyzing over 1 million individuals from five prospective cohort studies found that the supplementation with vitamin E supplements was associated with lower rates of ALS (Wang et al. 2011). Findings hinted towards vitamin E possibly having a protective effect, going along with the hypothesis that oxidative stress serves as a driving mechanism for ALS (Wang et al. 2011).

The supplementation of coenzyme Q10 (CoQ10), an antioxidant that plays a major role in the electron transport chain in the mitochondria, has also been investigated in patients with ALS. Early studies demonstrated significantly extended survival in mouse models of ALS following supplementation with CoQ10 (Beal 2002). However, one case-control study found no significant difference in the serum levels of CoQ10 between patients with ALS and matched controls, nor a correlation between the serum levels of CoQ10 and age of onset or progression of the disease

(Molina et al. 2000). A phase II trial also found no evidence to support CoQ10 supplementation in patients with ALS (Kaufmann et al. 2009).

One promising agent in the treatment of ALS is masitinib, a highly selective oral tyrosine kinase inhibitor that reduces central nervous system neuroinflammation (Paul et al. 2010). Masitinib does not target oxidative stress mechanisms or protein aggregates located in motor neurons, rather, it acts to quell the downstream effects of these processes and address neuroinflammation by targeting glial cells and inflammatory mediators (Trias et al. 2016). Studies performed on rodent models of ALS suggest that the progression of paralysis is mediated by the proliferation and activation of glial cells (Trias et al. 2016). Glial cells release inflammatory mediators while proliferating and surrounding degenerating motor neurons. Masitinib allows for the pharmacological downregulation of aberrant glial cells, reducing spinal cord neuroinflammation. A study conducted on rat models of ALS found that the administration of masitinib after the onset of paralysis prolonged survival by 40% (Trias et al. 2016). A recent randomized control trial found that masitinib was significantly more effective in patients with ALS and could potentially prolong patient survival by over two years as compared to placebo (Mora et al. 2020).

Cannabinoids, the bioactive compounds found in *Cannabis sativa,* are known to exhibit antioxidant, anti-inflammatory, and anti-excitotoxic effects (Giacoppo and Mazzon 2016). Several preclinical studies on transgenic ALS rodent models were conducted to assess the effects of cannabinoids on the progression of ALS. Urbi et al. provide a comprehensive literature review and meta-analysis of studies using murine models to investigate the effects of cannabinoids on disease progression (Urbi et al. 2019). Results showed cannabinoids to reduce elevated glutamate levels in the central nervous system, decrease neuroinflammation by inhibiting microglial activation, and inhibit the formation of ROS. To date, no clinical trials were performed regarding the therapeutic effects of cannabinoids in ALS. One observational study noted that patients with ALS using cannabis experienced decreased rates and levels of symptoms such as depression, pain, appetite loss, spasticity, and drooling (Amtmann et al. 2004).

More recently, attention has been given to a combined formulation of sodium phenylbutyrate (PB) and taurursodiol (TURSO), which reduces neuronal apoptosis by targeting mitochondrial dysfunction and endoplasmic reticulum stress (Paganoni et al. 2021). Several preclinical studies showed the mitigating effect of PB, TURSO, and combined PB-TURSO on neuronal death (Del Signore et al., 2009; Ryu et al. 2005). A recent randomized placebo-controlled trial conducted by Paganoni et al. found that patients treated with PB-TURSO experienced a median survival of 6.5 months longer as compared to placebo. In a separate study, PB-TURSO slowed the functional decline of patients with ALS for 24 weeks (Paganoni et al. 2022).

8.5 Conclusions and Future Directions

While it is known that oxidative stress is one of the major mechanisms involved in ALS pathogenesis, whether it serves as a primary cause or simply a downstream consequence of other pathological processes remains unknown. Several biomarkers of oxidative stress have been identified in patients with ALS, including 8-OHdG, 3-NT, HNE, MDA, and potentially SOD-1. Some of these might serve as potential biomarkers to diagnose the disease and monitor its progression. Conflicting findings, small sample sizes, and sparse research limit the current clinical utility of these biomarkers. FDA-approved medications for the treatment of ALS include riluzole and edaravone, which target neuronal excitotoxicity and oxidative stress mechanisms, respectively. Several preclinical experiments, clinical studies, and randomized controlled trials are looking into the role of molecules that target oxidative stress as potential interventions to stop the progression of – and possibly treat – ALS. Future research should further investigate the role of oxidative stress in the pathophysiology and progression of ALS. Identifying biomarkers that hint towards the illness and its progression is vital as it can ameliorate the quality of life and expectancy. Biomarkers should be easy to collect, affordable, and clinically sensitive. More clinical trials with adequate sample sizes are also necessary to identify potential therapeutic interventions.

References

Amtmann D, Weydt P, Johnson KL, Jensen MP, Carter GT (2004) Survey of cannabis use in patients with amyotrophic lateral sclerosis. Am J Hosp Palliat Care 21(2):95–104. https://doi.org/10.1177/104990910402100206

Ayers JI, Cashman NR (2018) Prion-like mechanisms in amyotrophic lateral sclerosis. Handb Clin Neurol 153:337–354. https://doi.org/10.1016/B978-0-444-63945-5.00018-0

Baillet A, Chanteperdrix V, Trocmé C, Casez P, Garrel C, Besson G (2010) The role of oxidative stress in amyotrophic lateral sclerosis and Parkinson's disease. Neurochem Res 35(10):1530–1537. https://doi.org/10.1007/s11064-010-0212-5. Epub 2010

Bandookwala M, Sengupta P (2020) 3-Nitrotyrosine: a versatile oxidative stress biomarker for major neurodegenerative diseases. Int J Neurosci 130(10):1047–1062. https://doi.org/10.1080/00207454.2020.1713776. Epub 2020

Beal MF (2002) Coenzyme Q10 as a possible treatment for neurodegenerative diseases. Free Radic Res 36(4):455–460. https://doi.org/10.1080/10715760290021315

Beal MF, Ferrante RJ, Browne SE, Matthews RT, Kowall NW, Brown RH Jr (1997) Increased 3-nitrotyrosine in both sporadic and familial amyotrophic lateral sclerosis. Ann Neurol 42(4):644–654. https://doi.org/10.1002/ana.410420416

Benkler C, O'Neil AL, Slepian S et al (2018) Aggregated SOD1 causes selective death of cultured human motor neurons. *Sci Rep* **8**:16393. https://doi.org/10.1038/s41598-018-34759-z

Berdyński M, Miszta P, Safranow K, Andersen PM, Morita M, Filipek S, Żekanowski C, Kuźma-Kozakiewicz M (2022) SOD1 mutations associated with amyotrophic lateral sclerosis analysis of variant severity. Sci Rep 12(1):103. https://doi.org/10.1038/s41598-021-03891-8. PMID: 34996976; PMCID: PMC8742055

Bhandari R, Kuhad A, Kuhad A (2018) Edaravone: a new hope for deadly amyotrophic lateral sclerosis. Drugs Today (Barc) 54(6):349–360. https://doi.org/10.1358/dot.2018.54.6.2828189

Blasco H, Garcon G, Patin F, Veyrat-Durebex C, Boyer J, Devos D, Vourc'h P, Andres CR, Corcia P (2017) Panel of oxidative stress and inflammatory biomarkers in ALS: a pilot study. Can J Neurol Sci 44(1):90–95. https://doi.org/10.1017/cjn.2016.284. Epub 2016

de Boer EMJ, Orie VK, Williams T et al (2021) TDP-43 proteinopathies: a new wave of neurodegenerative diseases. J Neurol Neurosurg Psychiatry 92:86–95

Bogdanov M, Brown RH, Matson W, Smart R, Hayden D, O'Donnell H, Flint Beal M, Cudkowicz M (2000) Increased oxidative damage to DNA in ALS patients. Free Radic Biol Med 29(7):652–658. https://doi.org/10.1016/s0891-5849(00)00349-x

Bond L, Bernhardt K, Madria P, Sorrentino K, Scelsi H, Mitchell CS (2018) A metadata analysis of oxidative stress etiology in preclinical amyotrophic lateral sclerosis: benefits of antioxidant therapy. Front Neurosci 12:10. https://doi.org/10.3389/fnins.2018.00010. PMID: 29416499; PMCID: PMC5787557

Bradley WG, Robison SH, Tandan R (1987) Deficient repair of alkylation damage of DNA in Alzheimer's disease and amyotrophic lateral sclerosis cells. Adv Exp Med Biol 209:3–6. https://doi.org/10.1007/978-1-4684-5302-7_1

Brizzi KT, Bridges JFP, Yersak J, Balas C, Thakur N, Galvin M, Hardiman O, Heatwole C, Ravits J, Simmons Z, Bruijn L, Chan J, Bedlack R, Berry JD (2020) Understanding the needs of people with ALS: a national survey of patients and caregivers. Amyotroph Lateral Scler Frontotemporal Degener 21(5-6):355–363. https://doi.org/10.1080/21678421.2020.1760889. Epub 2020

Cohen T, Hwang A, Restrepo C et al (2015) An acetylation switch controls TDP-43 function and aggregation propensity. *Nat Commun* **6**:5845. https://doi.org/10.1038/ncomms6845

Del Signore SJ, Amante DJ, Kim J, Stack EC, Goodrich S, Cormier K, Smith K, Cudkowicz ME, Ferrante RJ (2009) Combined riluzole and sodium phenylbutyrate therapy in transgenic amyotrophic lateral sclerosis mice. Amyotroph Lateral Scler 10(2):85–94. https://doi.org/10.1080/17482960802226148

Devos D, Moreau C, Kyheng M et al (2019) A ferroptosis–based panel of prognostic biomarkers for Amyotrophic Lateral Sclerosis. *Sci Rep* **9**:2918. https://doi.org/10.1038/s41598-019-39739-5

Fields CR, Bengoa-Vergniory N, Wade-Martins R (2019) Targeting alpha-synuclein as a therapy for Parkinson's disease. Front Mol Neurosci 12:299. https://doi.org/10.3389/fnmol.2019.00299

Fitzgerald KC, O'Reilly ÉJ, Fondell E, Falcone GJ, McCullough ML, Park Y, Kolonel LN, Ascherio A (2013) Intakes of vitamin C and carotenoids and risk of amyotrophic lateral sclerosis: pooled results from 5 cohort studies. Ann Neurol 73(2):236–245. https://doi.org/10.1002/ana.23820. Epub 2013. PMID: 23362045; PMCID: PMC3608702

Fujisawa A, Yamamoto Y (2016) Edaravone, a potent free radical scavenger, reacts with peroxynitrite to produce predominantly 4-NO-edaravone. Redox Rep 21(3):98–103. https://doi.org/10.1179/1351000215Y.0000000025. Epub 2016. PMID: 26196041; PMCID: PMC6837651

Gaber TA-ZK, Mehmood Z, Siringwani H (2016) Riluzole. Prog Neurol Psychiatry 20:32–33

Giacoppo S, Mazzon E (2016) Can cannabinoids be a potential therapeutic tool in amyotrophic lateral sclerosis? Neural Regen Res 11(12):1896–1899. https://doi.org/10.4103/1673-5374.197125. PMID: 28197175; PMCID: PMC5270417

Goh CW, Lee IC, Sundaram JR, George SE, Yusoff P, Brush MH, Sze NSK, Shenolikar S (2018) Chronic oxidative stress promotes GADD34-mediated phosphorylation of the TAR DNA-binding protein TDP-43, a modification linked to neurodegeneration. J Biol Chem 293(1):163–176. https://doi.org/10.1074/jbc.M117.814111. Epub 2017. PMID: 29109149; PMCID: PMC5766914

Hansson O (2021) Biomarkers for neurodegenerative diseases. *Nat Med* **27**:954–963. https://doi.org/10.1038/s41591-021-01382-x

Harley J, Clarke BE, Patani R (2021) The interplay of RNA binding proteins, oxidative stress and mitochondrial dysfunction in ALS. *Antioxidants* 10(4):552. https://doi.org/10.3390/antiox10040552

Hillen H (2019) The beta amyloid dysfunction (BAD) hypothesis for Alzheimer's disease. *Front Neurosci* 13:1154. https://doi.org/10.3389/fnins.2019.01154

Hinchcliffe M, Smith A (2017) Riluzole: real-world evidence supports significant extension of median survival times in patients with amyotrophic lateral sclerosis. Degener Neurol Neuromuscul Dis 7:61–70. https://doi.org/10.2147/DNND.S135748. PMID: 30050378; PMCID: PMC6053101

Hosaka T, Tsuji H, Tamaoka A (2021) Biomolecular modifications linked to oxidative stress in amyotrophic lateral sclerosis: determining promising biomarkers related to oxidative stress. *Processes* 9(9):1667. https://doi.org/10.3390/pr9091667

Houzen H, Kano T, Horiuchi K, Wakita M, Nagai A, Yabe I (2021) Improved long-term survival with Edaravone therapy in patients with Amyotrophic Lateral Sclerosis: a retrospective single-center study in Japan. Pharmaceuticals (Basel) 14(8):705. https://doi.org/10.3390/ph14080705. PMID: 34451802; PMCID: PMC8398196

Hulisz D (2018) Amyotrophic lateral sclerosis: disease state overview. *AJMC*, MJH Life Sciences, 23, https://www.ajmc.com/view/amyotrophic-lateral-sclerosis-disease-state-overview.

Ihara Y, Mori A, Hayabara T, Kawai M, Namba R, Nobukuni K, Sato K, Kibata M (1995) Superoxide dismutase and free radicals in sporadic amyotrophic lateral sclerosis: relationship to clinical data. J Neurol Sci 134(1-2):51–56. https://doi.org/10.1016/0022-510x(95)00193-6

Jagaraj CJ, Parakh S, Atkin JD (2021) Emerging evidence highlighting the importance of redox dysregulation in the pathogenesis of amyotrophic lateral sclerosis (ALS). Front Cell Neurosci 14:581950. https://doi.org/10.3389/fncel.2020.581950. PMID: 33679322; PMCID: PMC7929997

Kaufmann P, Thompson JL, Levy G, Buchsbaum R, Shefner J, Krivickas LS, Katz J, Rollins Y, Barohn RJ, Jackson CE, Tiryaki E, Lomen-Hoerth C, Armon C, Tandan R, Rudnicki SA, Rezania K, Sufit R, Pestronk A, Novella SP, Heiman-Patterson T, Kasarskis EJ, Pioro EP, Montes J, Arbing R, Vecchio D, Barsdorf A, Mitsumoto H, Levin B, QALS Study Group (2009) Phase II trial of CoQ10 for ALS finds insufficient evidence to justify phase III. Ann Neurol 66(2):235–244. https://doi.org/10.1002/ana.21743. PMID: 19743457; PMCID: PMC2854625

Kim BW, Jeong YE, Wong M et al (2020) DNA damage accumulates and responses are engaged in human ALS brain and spinal motor neurons and DNA repair is activatable in iPSC-derived motor neurons with SOD1 mutations. *Acta Neuropathol Commun* **8**:7. https://doi.org/10.1186/s40478-019-0874-4

King AE, Woodhouse A, Kirkcaldie MT, Vickers JC (2016) Excitotoxicity in ALS: overstimulation, or overreaction? Exp Neurol 275(Pt 1):162–171. https://doi.org/10.1016/j.expneurol.2015.09.019. Epub 2015

Kisby GE, Milne J, Sweatt C (1997) Evidence of reduced DNA repair in amyotrophic lateral sclerosis brain tissue. Neuroreport 8:1337–1339

Lévy E, El Banna N, Baïlle D, Heneman-Masurel A, Truchet S, Rezaei H, Huang ME, Béringue V, Martin D, Vernis L (2019) Causative links between protein aggregation and oxidative stress: a review. Int J Mol Sci 20(16):3896. https://doi.org/10.3390/ijms20163896. PMID: 31405050; PMCID: PMC6719959

Limanaqi F, Biagioni F, Mastroiacovo F, Polzella M, Lazzeri G, Fornai F (2020) Merging the multi-target effects of phytochemicals in neurodegeneration: from oxidative stress to protein aggregation and inflammation. *Antioxidants* 9(10):1022. https://doi.org/10.3390/antiox9101022

Longinetti E, Fang F (2019) Epidemiology of amyotrophic lateral sclerosis: an update of recent literature. Curr Opin Neurol 32(5):771–776. https://doi.org/10.1097/WCO.0000000000000730. PMID: 31361627; PMCID: PMC6735526

Martin LJ, Liu Z, Chen K, Price AC, Pan Y, Swaby JA, Golden WC (2007) Motor neuron degeneration in amyotrophic lateral sclerosis mutant superoxide dismutase-1 transgenic mice: mechanisms of mitochondriopathy and cell death. J Comp Neurol 500(1):20–46. https://doi.org/10.1002/cne.21160

Masrori P, Van Damme P (2020) Amyotrophic lateral sclerosis: a clinical review. Eur J Neurol 27(10):1918–1929. https://doi.org/10.1111/ene.14393. Epub 2020. PMID: 32526057; PMCID: PMC7540334

Miana-Mena FJ, González-Mingot C, Larrodé P et al (2011) Monitoring systemic oxidative stress in an animal model of amyotrophic lateral sclerosis. *J Neurol* **258**:762–769. https://doi.org/10.1007/s00415-010-5825-8

Milne GL, Dai Q, Roberts LJ 2nd. (2015) The isoprostanes--25 years later. Biochim Biophys Acta 1851(4):433–445. https://doi.org/10.1016/j.bbalip.2014.10.007. Epub 2014. PMID: 25449649; PMCID: PMC5404383

Mitsumoto H, Santella RM, Liu X, Bogdanov M, Zipprich J, Wu HC, Mahata J, Kilty M, Bednarz K, Bell D, Gordon PH, Hornig M, Mehrazin M, Naini A, Flint Beal M, Factor-Litvak P (2008) Oxidative stress biomarkers in sporadic ALS. Amyotroph Lateral Scler 9(3):177–183. https://doi.org/10.1080/17482960801933942. PMID: 18574762; PMCID: PMC4332387

Molina JA, de Bustos F, Jiménez-Jiménez FJ, Gómez-Escalonilla C, García-Redondo A, Esteban J, Guerrero-Sola A, del Hoyo P, Martínez-Salio A, Ramírez-Ramos C, Indurain GR, Arenas J (2000) Serum levels of coenzyme Q10 in patients with amyotrophic lateral sclerosis. J Neural Transm (Vienna) 107(8-9):1021–1026. https://doi.org/10.1007/s007020070050

Mora JS, Genge A, Chio A, Estol CJ, Chaverri D, Hernández M, Marín S, Mascias J, Rodriguez GE, Povedano M, Paipa A, Dominguez R, Gamez J, Salvado M, Lunetta C, Ballario C, Riva N, Mandrioli J, Moussy A, Kinet JP, Auclair C, Dubreuil P, Arnold V, Mansfield CD, Hermine O, AB10015 STUDY GROUP (2020) Masitinib as an add-on therapy to riluzole in patients with amyotrophic lateral sclerosis: a randomized clinical trial. Amyotroph Lateral Scler Frontotemporal Degener 21(1-2):5–14. https://doi.org/10.1080/21678421.2019.1632346. Epub 2019

Moumen R, Nouvelot A, Duval D, Lechevalier B, Viader F (1997) Plasma superoxide dismutase and glutathione peroxidase activity in sporadic amyotrophic lateral sclerosis. J Neurol Sci 151(1):35–39. https://doi.org/10.1016/s0022-510x(97)00109-3

Nightingale B (2020) The effect of edaravone on amyotrophic lateral sclerosis. *J Neurol Res* 10(5):150–159. [Online], Web. 4. 2022

Okada M, Yamashita S, Ueyama H, Ishizaki M, Maeda Y, Ando Y (2018) Long-term effects of edaravone on survival of patients with amyotrophic lateral sclerosis. eNeurological Sci 11:11–14. https://doi.org/10.1016/j.ensci.2018.05.001. PMID: 29928711; PMCID: PMC6006910

Oliveira TT, Coutinho LG, de Oliveira LOA, Timoteo ARS, Farias GC, Agnez-Lima LF (2022) APE1/Ref-1 role in inflammation and immune response. Front Immunol 13:793096. https://doi.org/10.3389/fimmu.2022.793096. PMID: 35296074; PMCID: PMC8918667

Padayatty SJ, Katz A, Wang Y, Eck P, Kwon O, Lee JH, Chen S, Corpe C, Dutta A, Dutta SK, Levine M (2003) Vitamin C as an antioxidant: evaluation of its role in disease prevention. J Am Coll Nutr 22(1):18–35. https://doi.org/10.1080/07315724.2003.10719272

Paganoni S, Hendrix S, Dickson SP, Knowlton N, Berry JD, Elliott MA, Maiser S, Karam C, Caress JB, Owegi MA, Quick A, Wymer J, Goutman SA, Heitzman D, Heiman-Patterson TD, Jackson C, Quinn C, Rothstein JD, Kasarskis EJ, Katz J, Jenkins L, Ladha SS, Miller TM, Scelsa SN, Vu TH, Fournier C, Johnson KM, Swenson A, Goyal N, Pattee GL, Babu S, Chase M, Dagostino D, Hall M, Kittle G, Eydinov M, Ostrow J, Pothier L, Randall R, Shefner JM, Sherman AV, Tustison E, Vigneswaran P, Yu H, Cohen J, Klee J, Tanzi R, Gilbert W, Yeramian P, Cudkowicz M (2022) Effect of sodium phenylbutyrate/taurursodiol on tracheostomy/ventilation-free survival and hospitalisation in amyotrophic lateral sclerosis: long-term results from the CENTAUR trial. J Neurol Neurosurg Psychiatry 93(8):871–875. https://doi.org/10.1136/jnnp-2022-329024. Epub ahead of print. PMID: 35577511; PMCID: PMC9304116

Paganoni S, Hendrix S, Dickson SP, Knowlton N, Macklin EA, Berry JD, Elliott MA, Maiser S, Karam C, Caress JB, Owegi MA, Quick A, Wymer J, Goutman SA, Heitzman D, Heiman-Patterson TD, Jackson CE, Quinn C, Rothstein JD, Kasarskis EJ, Katz J, Jenkins L, Ladha S, Miller TM, Scelsa SN, Vu TH, Fournier CN, Glass JD, Johnson KM, Swenson A, Goyal

NA, Pattee GL, Andres PL, Babu S, Chase M, Dagostino D, Hall M, Kittle G, Eydinov M, McGovern M, Ostrow J, Pothier L, Randall R, Shefner JM, Sherman AV, St Pierre ME, Tustison E, Vigneswaran P, Walker J, Yu H, Chan J, Wittes J, Yu ZF, Cohen J, Klee J, Leslie K, Tanzi RE, Gilbert W, Yeramian PD, Schoenfeld D, Cudkowicz ME (2021) Long-term survival of participants in the CENTAUR trial of sodium phenylbutyrate-taurursodiol in amyotrophic lateral sclerosis. Muscle Nerve 63(1):31–39. https://doi.org/10.1002/mus.27091. Epub 2020. PMID: 33063909; PMCID: PMC7820979

Paul C, Sans B, Suarez F, Casassus P, Barete S, Lanternier F, Grandpeix-Guyodo C, Dubreuil P, Palmérini F, Mansfield CD, Gineste P, Moussy A, Hermine O, Lortholary O (2010) Masitinib for the treatment of systemic and cutaneous mastocytosis with handicap: a phase 2a study. Am J Hematol 85(12):921–925. https://doi.org/10.1002/ajh.21894

Petrov D, Mansfield C, Moussy A, Hermine O (2017) ALS clinical trials review: 20 years of failure. are we any closer to registering a new treatment? Front Aging Neurosci 9:68. https://doi.org/10.3389/fnagi.2017.00068. PMID: 28382000; PMCID: PMC5360725

Prasad A, Bharathi V, Sivalingam V, Girdhar A, Patel BK (2019) Molecular mechanisms of TDP-43 misfolding and pathology in amyotrophic lateral sclerosis. *Front Mol Neurosci* 12:25. https://doi.org/10.3389/fnmol.2019.00025

Qing X, Shi D, Lv X et al (2019) Prognostic significance of 8-hydroxy-2′-deoxyguanosine in solid tumors: a meta-analysis. *BMC Cancer* **19**:997. https://doi.org/10.1186/s12885-019-6189-9

Rosen D, Siddique T, Patterson D et al (1993) Mutations in Cu/Zn superoxide dismutase gene are associated with familial amyotrophic lateral sclerosis. *Nature* **362**:59–62. https://doi.org/10.1038/362059a0

Ryu H, Smith K, Camelo SI, Carreras I, Lee J, Iglesias AH, Dangond F, Cormier KA, Cudkowicz ME, Brown RH Jr, Ferrante RJ (2005) Sodium phenylbutyrate prolongs survival and regulates expression of anti-apoptotic genes in transgenic amyotrophic lateral sclerosis mice. J Neurochem 93(5):1087–1098. https://doi.org/10.1111/j.1471-4159.2005.03077.x. Erratum in: J Neurochem. 2006;96(3):908

Shaikh AY, Martin LJ (2002) DNA base-excision repair enzyme apurinic/apyrimidinic endonuclease/redox factor-1 is increased and competent in the brain and spinal cord of individuals with amyotrophic lateral sclerosis. Neuromolecular Med 2(1):47–60. https://doi.org/10.1007/s12017-002-0038-7

Singh A, Kukreti R, Saso L, Kukreti S (2019) Oxidative stress: a key modulator in neurodegenerative diseases. Molecules 24(8):1583. https://doi.org/10.3390/molecules24081583. PMID: 31013638; PMCID: PMC6514564

Singh Z et al (2014) Use of malondialdehyde as a biomarker for assessing oxidative stress in different disease pathologies: a review. *Iran J Public Health* 43(Supple 3):7–16

Suk TR, Rousseaux MWC (2020) The role of TDP-43 mislocalization in amyotrophic lateral sclerosis. *Mol Neurodegeneration* **15**:45. https://doi.org/10.1186/s13024-020-00397-1

Sun J, Carrero JJ, Zagai U, Evans M, Ingre C, Pawitan Y, Fang F (2020) Blood biomarkers and prognosis of amyotrophic lateral sclerosis. Eur J Neurol 27(11):2125–2133. https://doi.org/10.1111/ene.14409. Epub 2020

Tandan R, Robison SH, Munzer JS, Bradley WG (1987) Deficient DNA repair in amyotrophic lateral sclerosis cells. J Neurol Sci 79(1-2):189–203. https://doi.org/10.1016/0022-510x(87)90272-3

Tarafdar A, Pula G (2018) The role of NADPH oxidases and oxidative stress in neurodegenerative disorders. Int J Mol Sci 19(12):3824. https://doi.org/10.3390/ijms19123824. PMID: 30513656; PMCID: PMC6321244

Thakur S, Sarkar B, Cholia RP, Gautam N, Dhiman M, Mantha AK (2014) APE1/Ref-1 as an emerging therapeutic target for various human diseases: phytochemical modulation of its functions. Exp Mol Med 46(7):e106. https://doi.org/10.1038/emm.2014.42. PMID: 25033834; PMCID: PMC4119211

Tohgi H, Abe T, Yamazaki K, Murata T, Ishizaki E, Isobe C (1999) Remarkable increase in cerebrospinal fluid 3-nitrotyrosine in patients with sporadic amyotrophic lateral sclerosis. Ann Neurol 46(1):129–131. https://doi.org/10.1002/1531-8249(199907)46:1<129::aid-ana21>3.0.co;2-y

Tokuda E, Takei YI, Ohara S, Fujiwara N, Hozumi I, Furukawa Y (2019) Wild-type Cu/Zn-superoxide dismutase is misfolded in cerebrospinal fluid of sporadic amyotrophic lateral sclerosis. Mol Neurodegener 14(1):42. https://doi.org/10.1186/s13024-019-0341-5. PMID: 31744522; PMCID: PMC6862823

Trias E, Ibarburu S, Barreto-Núñez R, Babdor J, Maciel TT, Guillo M, Gros L, Dubreuil P, Díaz-Amarilla P, Cassina P, Martínez-Palma L, Moura IC, Beckman JS, Hermine O, Barbeito L (2016) Post-paralysis tyrosine kinase inhibition with masitinib abrogates neuroinflammation and slows disease progression in inherited amyotrophic lateral sclerosis. J Neuroinflammation 13(1):177. https://doi.org/10.1186/s12974-016-0620-9. PMID: 27400786; PMCID: PMC4940876

Urbi B, Owusu MA, Hughes I, Katz M, Broadley S, Sabet A (2019) Effects of cannabinoids in Amyotrophic Lateral Sclerosis (ALS) murine models: a systematic review and meta-analysis. J Neurochem 149(2):284–297. https://doi.org/10.1111/jnc.14639. Epub 2019

Wang H, O'Reilly ÉJ, Weisskopf MG, Logroscino G, McCullough ML, Schatzkin A, Kolonel LN, Ascherio A (2011) Vitamin E intake and risk of amyotrophic lateral sclerosis: a pooled analysis of data from 5 prospective cohort studies. Am J Epidemiol 173(6):595–602. https://doi.org/10.1093/aje/kwq416. Epub 2011. PMID: 21335424; PMCID: PMC3105261

Yoshino H, Kimura A (2006) Investigation of the therapeutic effects of edaravone, a free radical scavenger, on amyotrophic lateral sclerosis (phase II study) Amyotroph. Lateral Scler 7:241–245

MIX
Papier aus verantwortungsvollen Quellen
Paper from responsible sources
FSC® C105338

If you have any concerns about our products,
you can contact us on
ProductSafety@springernature.com

In case Publisher is established outside the EU,
the EU authorized representative is:
Springer Nature Customer Service Center GmbH
Europaplatz 3, 69115 Heidelberg, Germany

Printed by Libri Plureos GmbH
in Hamburg, Germany